The T-Cell Receptors

The T-Cell Receptors

Edited by

Tak W. Mak

The Ontario Cancer Institute
and Departments of Medical Biophysics and Immunology
University of Toronto
Toronto, Ontario, Canada

Plenum Press • New York and London

Library of Congress Cataloging in Publication Data

The T-cell receptors / edited by Tak W. Mak.
    p.   cm.
    Includes bibliographies and index.
    ISBN 0-306-42708-7
    1. T cells—Receptors. I. Mak, Tak W., date. [DNLM: 1. Antigen-Antibody Reactions. 2. Immunity, Cellular. 3. Lymphocytes—immunology. 4. Receptors, Antigen, T-Cell. QW 568 T113]
QR185.8.T2T33 1988
599′.029—dc 19

87-38495
CIP

© 1988 Plenum Press, New York
A Division of Plenum Publishing Corporation
233 Spring Street, New York, N.Y. 10013

Printed in the United States of America

Contributors

Balbino Alarcon, Laboratory of Molecular Immunology, Dana Farber Cancer Institute, Harvard Medical School, Boston, Massachusetts 02215

A. Neil Barclay, MRC Cellular Immunology Research Unit, Sir William Dunn School of Pathology, University of Oxford, Oxford OX1 3RE, England.

Mark A. Behlke, Howard Hughes Medical Institute, Departments of Medicine, Microbiology, and Immunology, Washington University, School of Medicine, St. Louis, Missouri 63110

Benjamin Berkhout, Laboratory of Molecular Immunology, Dana Farber Cancer Institute, Harvard Medical School, Boston, Massachusetts 02215

Nicolette Caccia, Department of Medicine and Medical Biophysics, Ontario Cancer Institute, University of Toronto, Toronto, Ontario M4X 1K9, Canada

Hubert S. Chou, Howard Hughes Medical Institute, Departments of Medicine, Microbiology, and Immunology, Washington University, School of Medicine, St. Louis, Missouri 63110

Hans Clevers, Laboratory of Molecular Immunology, Dana Farber Cancer Institute, Harvard Medical School, Boston, Massachusetts 02215

Katia Georgopoulos, Laboratory of Molecular Immunology, Dana Farber Cancer Institute, Harvard Medical School, Boston, Massachusetts 02215

Daniel Gold, Laboratory of Molecular Immunology, Dana Farber Cancer Institute, Harvard Medical School, Boston, Massachusetts 02215

H. Griesser, University of Toronto, Department of Medicine and Medical Biophysics, Ontario Cancer Institute, Toronto, Ontario M4X 1K9, Canada.

Gregory F. Hollis, Medical Oncology Branch, National Cancer Institute, National Institutes of Health, Bethesda, Maryland 20892

John Imboden, Howard Hughes Medical Institute, and Department of Medicine, University of California, San Francisco, California 94143. *Present address*: Immunology/Arthritis Section, Veterans Administration Medical Center, San Francisco, California 94121

Pauline Johnson, MRC Cellular Immunology Research Unit, Sir William Dunn School of Pathology, University of Oxford, Oxford OX1 3RE, England. *Present address:* Cancer Biology Lab, The Salk Institute, La Jolla, California 92112

Nobuhiro Kimura, Department of Medicine and Medical Biophysics, Ontario Cancer Institute, University of Toronto, Toronto, Ontario M4X 1K9, Canada

Ilan R. Kirsch, Medical Oncology Branch, National Cancer Institute, National Institutes of Health, Bethesda, Maryland 20892

Dennis Y. Loh, Howard Hughes Medical Institute, Departments of Medicine, Microbiology, and Immunology, School of Medicine, Washington University, St. Louis, Missouri 63110

Tak W. Mak, Department of Medicine and Medical Biophysics, Ontario Cancer Institute, University of Toronto, Toronto, Ontario M4X 1K9, Canada

Bernard Manger, Howard Hughes Medical Institute, and Department of Medicine, University of California, San Francisco, California 94143. *Present address:* Institute of Clinical Immunology, Medical School Erlangen, D-8529 Erlangen, Federal Republic of Germany

Geoff W. McCaughan, MRC Cellular Immunology Research Unit, Sir William Dunn School of Pathology, University of Oxford, Oxford OX1 3RE, England. *Present address:* A. W. Morrow Gastroenterology and Liver Centre, Royal Prince Alfred Hospital, Camperdown, Sydney, Australia

M. D. Minden, Department of Medicine and Medical Biophysics, Ontario Cancer Institute, University of Toronto, Toronto, Ontario M4X 1K9, Canada

Hans Oettgen, Laboratory of Molecular Immunology, Dana Farber Cancer Institute, Harvard Medical School, Boston, Massachusetts 02215

Carolyn Pettey, Laboratory of Molecular Immunology, Dana Farber Cancer Institute, Harvard Medical School, Boston, Massachusetts 02215

J. Slingerland, Department of Medicine and Medical Biophysics, Ontario Cancer Institute, University of Toronto, Toronto, Ontario M4X 1K9, Canada

Rosanne Spolski, Department of Medicine and Medical Biophysics, Ontario Cancer Institute, University of Toronto, Toronto, Ontario M4X 1K9, Canada

Yoshihiro Takihara, Department of Medicine and Medical Biophysics, Ontario Cancer Institute, University of Toronto, Toronto, Ontario M4X 1K9, Canada

Cox Terhorst, Laboratory of Molecular Immunology, Dana Farber Cancer Institute, Harvard Medical School, Boston, Massachusetts 02215

Barry Toyonaga, Department of Medicine and Medical Biophysics, Ontario Cancer Institute, University of Toronto, Toronto, Ontario M4X 1K9, Canada

Peter van Den Elsen, Laboratory of Molecular Immunology, Dana Farber Cancer Institute, Harvard Medical School, Boston, Massachusetts 02215

Arthur Weiss, Howard Hughes Medical Institute, and Department of Medicine, University of California, San Francisco, California 94143

Tom Wileman, Laboratory of Molecular Immunology, Dana Farber Cancer Institute, Harvard Medical School, Boston, Massachusetts 02215

Alan F. Williams, MRC Cellular Immunology Research Unit, Sir William
    Dunn School of Pathology, University of Oxford, Oxford OX1 3RE,
    England
Rolf M. Zinkernagel, Institute for Experimental Pathology, University Hospi-
    tal Zurich, CH-8091 Zurich, Switzerland

Preface

The importance of thymus-dependent cells, or T cells, in the generation of a successful immune response was first realized in the early sixties. In the following two decades, a succession of elegant experiments established the antigen specificity of T cells and their ability to perform both as regulatory and effector cells. T cells were shown to be essential in most immune reactions, playing a crucial role in augmenting the activity of effector T and B cells against 'foreign' antigen, as well as in the suppression of effector activity against self antigens. The means by which T cells differentiate 'foreign' from 'self' antigens is based on their recognition of antigen almost exclusively in the context of self major histocompatibility complex products, unlike B cells, which recognize antigen alone. It is this recognition, mediated by the T-cell receptor, that sets into motion the diverse cell–cell interactions, which control the differentiation and regulation of the immune response.

Although its importance was well established, the molecular nature of the T-cell receptor remained elusive for two decades. Many hypotheses as to its structure and precise function were put forward, using immunoglobulin as a basis for conjecture, but "the Holy Grail of Immunology" remained ephemeral until three years ago. In the ensuing years, both immunologists and molecular biologists have contributed to an explosion of data unsurpassed by any previous period in the field.

This book is a collection of articles by major contributors to the field, summarizing our present knowledge of the receptor and the genes that encode it. The first part of the book describes the structure and function of the T-cell receptor, including its place in the immunoglobulin supergene family, the structure of the genes encoding the T-cell receptor heterodimer and the T3 proteins with which it is associated, the role of the major histocompatibility complex (MHC) products in T-cell recognition, and the biochemical events following receptor-mediated activation of T lymphocytes. The repertoire and potential diversity of the T-cell receptor in man and mouse and the rearrangement of the genes encoding the receptor during T-cell ontogeny in the thymus are also described. The latter half of the book deals with the use of T-cell receptor genes to investigate some interesting areas of immunology such as the repertoire of receptor genes in mutant mice, the role of the T-cell receptor in MHC-related disease, and the involvement of the T-cell receptor loci in chromosomal abnormalities characteristic of T-cell malignancies. Other chapters describe the use of these genes in the definition of the clonality and lineage of malignant and proliferative T-cell diseases, and as probes for the identification and isolation of oncogenes involved in the generation of T-cell-related malignancies. The last chapter deals with the γδ heterodimer, another T-cell receptor, which is expressed on certain, T-cell subpopulations.

These chapters provide a synthesis of the large volume of data and the emerging concepts that lay a foundation for the resolution of some of the most fascinating puzzles facing immunologists today, including the relationship between the T-cell receptor genes and autoimmune disease, and the fundamental questions of how tolerance to self-antigens is acquired and how T cells learn to recognize antigens in the context of self MHC products.

I would like to express my appreciation to N. Caccia and D. Quon for assisting in the coordination of this book.

Tak W. Mak

Toronto

Contents

Chapter 4

T-Cell Receptor Genes: Mutant Mice and Genes   89

Dennis Y. Loh, Mark A. Behlke, and Hubert S. Chou

Chapter 5

Thymic Ontogeny and the T-Cell Receptor Genes 101

Nicolette Caccia, Rosanne Spolski, and Tak W. Mak

Chapter 6

**The T-Cell Receptor/T3 Complex on the
Surface of Human and Murine T Lymphocytes** 117

Cox Terhorst, Benjamin Berkhout, Balbino Alarcon,
Hans Clevers, Katia Georgopoulos, Daniel Gold, Hans Oettgen,
Carolyn Pettey, Peter van Den Elsen, and Tom Wileman

Chapter 7

**Role of the T3/T-Cell Antigen
Receptor Complex in T-Cell Activation** 133

Bernard Manger, John Imboden, and Arthur Weiss

Chapter 8

**The Structure and Expression of the T-Cell α-, β-,
and γ-Chain Genes in Human Malignancies** 151

J. Slingerland, H. Griesser, T. W. Mak, and M. D. Minden

Chapter 9

**The Involvement of the T-Cell Receptor
in Chromosomal Aberrations** 175

Ilan R. Kirsch and Gregory F. Hollis

Chapter 10

MHC-Disease Associations and
T-Cell-Mediated Immunopathology 195

Rolf M. Zinkernagel

Chapter 11

The γ—δ Heterodimer: A Second T-Cell Receptor? 205

Nicolette Caccia, Yoshihiro Takihara, and Tak W. Mak

The T-Cell Receptors

1

Introduction

NICOLETTE CACCIA and TAK W. MAK

The regulation of the immune response is a complex process that involves a number of different cells and their products. Recent investigations of both auto-immune disorders and MHC-linked diseases have highlighted the important role that T cells play in these diseases and, thus their role in the maintenance of a well-balanced immune response crucial to the health of an organism.

T cells have a diverse, clonally distributed repertoire and recognize antigen in a specific manner, but this recognition is unique in that antigen can only be recognized on a cell surface and only in the context of major histocompatibility complex (MHC) products, a phenomenon known as MHC restriction. It is this restriction that enables T cells to distinguish 'self' antigens from 'foreign' molecules and thus to aid in the regulation of the response against foreign antigens, while preventing a response against self. The recognition of antigen is mediated by the T-cell receptor and in this book we have tried to summarize the present understanding of the receptor in the context of T-cell activation and function. We have divided the book into two parts: the first of which describes the structure and function of the T-cell receptor and the genes that encode it, while the second outlines the role that this receptor plays in a number of interesting immunological problems.

An understanding of the T-cell receptor remained an elusive goal for a number of years and with the recent cloning of the genes encoding it, an explosion of information was generated by the many interested investigators. The receptor itself is a cell surface heterodimer composed of an acidic α chain and a more basic β chain. Both chains contain variable and constant domains and the genes encoding these chains are composed of separate, noncontiguous gene seg-

NICOLETTE CACCIA and TAK W. MAK • Department of Medicine and Medical Biophysics, Ontario Cancer Institute, University of Toronto, Toronto, Ontario M4X 1K9, Canada.

ments in the germline, which rearrange to produce a functional gene, allowing for the generation of a wide variety of receptors by the combinatorial use of these segments. Chapter 2 outlines the structure of the α and β chain genes, the rearrangement process involved in the generation of a diverse repertoire, and the role that these chains play in the recognition of antigen and MHC.

In the search for the receptor genes, a number of other T-cell specific molecules were isolated. One of the most interesting of these is the γ chain, a cell surface molecule that is homologous to the α and β chains and rearranges in T cells. The TcRα, TcRβ, and γ chains are all members of the immunoglobulin gene superfamily, which has taken shape over the past few years. Its members have diverse functions and expression patterns in various cell types and are, for the most part, unlinked in the genome, yet they share a number of characteristics, in addition to their structural similarities. The members of this superfamily are cell surface molecules with no known enzymatic activity, whose known functions include the binding to other molecules and the subsequent triggering of cellular events.

In Chapter 3, Barclay *et al.* describe the members of this superfamily, which includes molecules that may not be involved in the recognition of antigen, such as MRC OX-2, and those which are not apparently immune-system specific, such as Thy 1, as well as the more traditional proteins, which are involved in the binding of antigen (immunoglobulin and the T-cell receptor) or participate in immune interactions [MHC gene products, CD4 (Leu3/L3T4), CD2, CD1, CD5, and CD8 (Leu2a/Lyt2)]. All the members have regions of similarity that are about 100 amino acids long, have cysteine residues in similar positions, and seem to have similar β-strand folding patterns in the sequences between the cysteines, suggesting a common primordial ancestor with these features. These regions, or domains, can be classified as C-like or V-like based on the number of β-loops, with the V-like structures possessing an extra loop in the middle of the domain.

Barclay *et al.* propose a scheme for the evolution of this superfamily, based on homology between members, which suggests that their ancestral gene mediated recognition between cells that were not involved in immunity. They propose that immune recognition arose from a modification of a recognition system involved in programmed death of a given cell by another cell, in the course of cellular differentiation. It is interesting to note that this kind of programmed death is important in neural tissue development and that this hypothesis would explain the prevalence of molecules belonging to the immunoglobulin supergene family on the surface of brain cells. This scheme also proposes that the MHC genes were the first genes involved in immune recognition to arise, followed by T-cell receptor genes, and then by immunoglobulin genes, which is the order of appearance of these genes as one ascends the evolutionary ladder of organisms from coelenterates to mammals.

One of the more interesting phenomena that one sees when comparing immune system genes from different species is the variation in the numbers of germline sequences from which an organism can draw to produce a sufficient repertoire of antigen-specific molecules to deal with the daily onslaught of foreign molecules. Examination of the number of V_β genes in different murine strains provides an interesting example of this variation. Certain laboratory mouse strains have lost a significant proportion of their germline V_β gene segments. In SJL mice, ten V_β from six subfamilies have been deleted from the genome, and have apparently not been replaced by new V_β sequences. These mice, however, seem to have an adequate immune response, implying that a large number of V_β genes may not be necessary.

In Chapter 4, Loh *et al.* discuss the T-cell receptor genes in mutant mice, including the differences in variable gene repertoire and the effects of changes in repertoire size on the immune response. They also explore possible interactions between the T-cell receptor and MHC genes, and the connection between T-cell receptor repertoire and MHC-linked disease.

Despite the fact that the thymus plays an important role in the selection and education of T cells, this organ remains very much a black box. T-cell precursors from the fetal liver, and in more mature animals, the bone marrow, migrate to the thymus where they proliferate and differentiate into functional T cells. One of the critical stages in the differentiation process involves the rearrangement of T-cell receptor genes to produce a functional cell surface molecule. Since it is this molecule that provides recognition of antigen and MHC, its expression enables the positive selection of cells bearing appropriate specificities and the negative selection of cells specific for 'self' antigens. The rearrangement and expression of TcR genes in different thymocyte subpopulations and the various differentiation pathways within the embryonic and adult thymus are presented in Chapter 5. The different maturation schemes proposed by different groups and the place that TcRα, TcRβ, and γ and δ gene rearrangement and expression take in the evolution of a coherent model of thymic development are also discussed.

In both the thymus and the periphery, the T-cell receptor heterodimer is closely associated with the T3 proteins on the cell surface, and it would seem that while the heterodimer provides the recognition function, it is T3 that plays a role in transmembrane signaling events. In Chapter 6, Terhorst *et al.* describe the component molecules of the T3 and their association with the T-cell receptor heterodimer.

Activation of T cells, upon the recognition of antigen and MHC gene products, gives rise to a number of events, including the production of lymphokines that regulate the immune response, the appearance of new cell-surface markers and the proliferation of T cells. The T-cell receptor complex mediates this activation by the recognition of antigen in the context of self MHC and the conversion of this specific recognition event into a transmembrane signal that initiates T-cell activation. The activation process would seem to be the same in all T cells, but it has only been recently that the specific events that occur upon the binding of antigen by T cells have come to light. Manger *et al.* describe in Chapter 7 the role that the T-cell receptor complex plays in the initiation of activation and outline the investigation of the early events of this process.

Activation of a resting T cell is a two-signal event. In addition to the recognition of immobilized antigen in the context of self MHC, usually on the surface of an antigen presenting (APC), a second signal, which is usually provided by IL-1 secreted by the APC, is needed. The binding of antigen can be mimicked by T-cell receptor complex-specific antibodies and the effect of IL-1 by compounds such as PMA, allowing the dissection of activation events.

The binding of the receptor complex initiates polyphosphoinositide hydrolysis, leading to the generation of inositol triphosphate (IP_3) and diacylglycerol (DG). IP_3 induces the release of Ca^{2+} from intracellular stores, leading to an increase in the concentration of intracellular Ca^{2+}, while DG translocates protein kinase C (PKC) to the membrane from the cytoplasm, thereby activating it.

It is this PKC translocation that is effected by PMA, but unlike PMA, DG is rapidly metabolized within the cell leading to only a transient increase in PKC activity. If the moiety binding the T-cell receptor complex is immobilized, as is antigen on APC, the increase in PKC activity continues for a longer period of time, which is sufficient to restimulate activated T cells, perhaps by allowing for the phosphorylation of specific substrates necessary for further intracellular events of T-cell activation. However, activation of resting T cells requires a second signal, such as that delivered by the binding of IL-1.

The increase in Ca^{2+} concentration and activation of PKC are intracellular activation signals common to a variety of cell types. Manger *et al.* describe in detail the early events of T-cell activation and the means by which they were divined, providing a stepping stone for interesting avenues of investigation.

One of the more practical uses for the T-cell receptor genes is in the classification and management of patients with T-cell malignancies. Together with the immunoglobulin genes, the T-cell receptor genes can be used to diagnose a disease as one of B- or T-cell origin, and then can be used as probes to determine if the malignancy is clonal and to follow the course of the disease by monitoring changes in this clonality. These changes can be used to assess the effects of a chemotherapeutic program and to detect relapses earlier, as well as to identify the progression of the disease, in terms of the emergence of new clonal populations and other changes in tumor composition.

In Chapter 8, Slingerland *et al.* discuss the use of T-cell receptor probes in the diagnosis and management of T-cell disorders, as well as, in the study of chromosomal translocations characteristic of certain T-cell malignancies.

Over the past few decades, an increasing number of distinctive chromosomal abnormalities have become associated with certain neoplasias. The fact that these abnormalities are found only in the tumor cells, not in the nonmalignant cells, of an otherwise karyotypically normal person, led to the proposal that these abnormalities are involved in tumorigenesis. One of the best-characterized abnormalities is the translocation between the locus encoding the cellular proto-oncogene, c-*myc,* and the immunoglobulin loci, in the human and murine B cell tumors, Burkitt's lymphoma, and murine plasmacytoma. This connection between expression of cell-type specific genes and their involvement in chromosomal abnormalities led to the proposal that the T-cell receptor genes would be located at the loci, which are consistently involved in translocations in T-cell tumors. This hypothesis was borne out by the mapping of the α chain to chromosome 14 and the β chain to chromosome 7 in man, both of which are involved in abnormalities in T-cell disorders.

In Chapter 9, Kirsch and Hollis report the latest findings about these T-cells chromosomal abnormalities and compare them to the Burkitt's lymphoma model. Inversion and translocation between the human immunoglobulin heavy chain and T-cell receptor α chain loci on chromosome 14 are two common abnormalities in T-cell lymphomas, and since the same recombinase system seems to be used for rearrangement of both loci, it is likely that this recombination system is involved in the translocation process.

The study of T-cell specific tumors may aid in the understanding of B-cell specific tumors and vice versa. Kirsch and Hollis describe the similarities and differences between the two tumor types and propose a number of hypotheses as to their genesis and the possible involvement of as-yet-unknown cellular proto-oncogenes.

Another interesting association between the T-cell receptor and disease involves the link between certain MHC haplotypes and a predisposition to given chronic diseases. In some cases, such as human ankylosing spondylitis, this association is so strong that the final diagnosis is based upon HLA haplotyping of the patient, but in the majority of cases, the link is weak. A weak correlation between a given MHC haplotype and a certain disease may be the result of a number of complicating factors, such as the mediating effect of other loci or the antigenic variabilities of both the infectious agent and the self MHC molecules with which they are associated. In some instances an MHC-linked disease may be the result of an inherent defect in the MHC antigens or in certain closely linked genes, such as those encoding the complement proteins or 21-hydroxylase. However, in a large proportion of the diseases where the link is not strong, it would seem that T-cell response plays a role in their pathogenesis. Given that T cells are instrumental in the regulation of the immune response, and their activation is dependent on their recognition of antigen in the context of self MHC, their involvement could take one of two forms. T cells may not be able to respond to a given antigen, thus allowing a cytopathic agent free rein, or an agent, which may or may not be cytopathic, by virtue of its infection of a certain cell type may indirectly trigger a MHC-associated T-cell-mediated response to cells of that type. In both cases the infectious agent would have to be poorly cytopathic or a cytopathic agent that has arisen recently for evolution to have tolerated the lack of response in the first case or the chronic low-grade response/low grade infection state in the second.

In Chapter 10, Zinkernagel explores the T-cell mediated immunopathology characteristic of the second type of MHC-related diseases. He proposes that there are three possible host responses to infection. There must be a strong response to acute, strongly cytopathic agents to rid the organism of the infection and ensure reproductive survival. Agents that are poorly cytopathic or not cytopathic can either trigger a weak response or no response, leading to chronic disease by the establishment of a balanced host–parasite state. A weak T-cell response may result in chronic infection and slow immunopathology caused by T-cell mediated tissue damage or the triggering of autoantibodies, while the absence of response will give an asymptomatic carrier state, if the agent is non-cytopathic, or low grade disease, if it is poorly cytopathic. Zinkernagel concentrates on the results of a weak T-cell response, using LCMV-infected mice as a model. He describes the effect of MHC haplotype on LCMV susceptibility and the pathology in strains with different haplotypes. He then carries his conclusions to the realm of human disease, drawing comparisons to the LCMV mouse model. He suggests that while it may be possible to prevent some of these diseases with an appropriate vaccine, the vaccination may alter the host–parasite balance, such that those who were previously asymptomatic carriers become the new population affected by the disease.

Chapter 11 outlines our present knowledge of the γ δ receptor. This second T-cell receptor shares many of the characteristics of the α β receptor, but seems to play a very different role in T cells. Like the α and β chains, both γ and δ are encoded by noncontiguous gene segments in the genome, but their diversity of genomic sequences is much less. There is a high proportion of nonfunctional γ-chain rearrangements in both thymocytes and peripheral blood T cells, and the γ δ receptor is found only on cells from certain specialized subpopulations of T cells, which do not have functional α β receptors, suggesting that γ δ may perform a function very different from that of the traditional α β receptor. Studies of the specificity of γ δ-bearing cells and of their functional activity promise to provide clues as to the nature of this second T-cell receptor.

The isolation of the T-cell receptor genes has permitted novel approaches for the investigation of certain immunological problems. We have tried to bring together the results of some of these investigations against a background of the information gathered about the α-, β-, and γ-chain genes. We hope that these chapters will be of use to both immunologists and interested people from other disciplines. There are still a number of interesting problems in immunology which T-cell receptor genes can be used to tackle, such as the molecular basis of T-cell recognition of MHC and antigen, the mechanisms of thymic selection, and the regulation of the immune response by T cells. Finally, we hope that this book will be followed by others, detailing the answers to these questions.

2

The α and β Chains of the T-Cell Receptor

NICOLETTE CACCIA, BARRY TOYONAGA, NOBUHIRO KIMURA, and TAK W. MAK

1. Introduction

The mammalian immune system must be equipped to recognize and eliminate a vast number of different foreign antigens. These antigens must be distinguished from self molecules to ensure that the organism's own structures are preserved, necessitating a well-regulated network of cells and their products. The immune system can be divided into the closely interacting compartments of nonspecific immunity and specific immunity. Nonspecific immunity is effected by cells such as macrophages and natural killer cells, which eliminate foreign antigen by lytic and digestive processes. These cells, by themselves, cannot distinguish between foreign and self antigens because they are nonspecific, and thus must be closely regulated and directed. This regulation is effected by cells from the specific immune compartment, B and T lymphocytes, which provide the fine tuning of the immune response. These two cell types are very similar in their specific recognition of antigen by means of a diverse clonally distributed repertoire, but differ in the molecules used to mediate their recognition of antigen and the context in which it is recognized. B lymphocytes secrete immunoglobulins and express them on their cell surfaces. These molecules are capable of recognizing both soluble antigen and that on the surface of cells. The portion of the molecule that is not involved in antigen recognition participates in a number of immune reactions, including those involving cells of the nonspecific compartment, that lead to the elimination of antigen (Davies and Metzger, 1983).

NICOLETTE CACCIA, BARRY TOYONAGA, NOBUHIRO KIMURA, and TAK W. MAK • Department of Medicine and Medical Biophysics, Ontario Cancer Institute, University of Toronto, Toronto, Ontario M4X 1K9, Canada.

There are several classes of T lymphocytes that play diverse roles in the immune system. Cytotoxic T lymphocytes (CTL), kill cells that display foreign antigen, and usually recognize the antigen in the context of MHC class I molecules. Helper T lymphocytes (HTL) augment T- and B-cell responses and tend to recognize Ag in the context of class II MHC molecules. Despite their different functions, it would seem that the recognition of antigen by helper and cytotoxic cells can be ascribed to the same structure, the T-cell receptor (TcR), which is distinct from immunoglobulin. Unlike B cells, T cells can only recognize antigen if it is on the surface of a cell, and then only if it is in the context of a self major histocompatibility complex (MHC) product, a phenomenon known as MHC restriction (Zinkernagel and Doherty, 1975). Thus, the differences in antigen recognition between B and T cells are likely to be reflected in differences between the T-cell receptor and immunoglobulin.

Over the past few decades immunoglobulin has been characterized extensively, but the nature of the T-cell receptor remained elusive until recently. With the introduction of long-term culture of T cells, clonal populations of T cells could be analysed and their receptor function investigated. Monoclonal antibodies, raised against idiotype-bearing T cell-specific cell surface proteins, could prevent the binding of T-cell targets and mimic antigen-specific activation, suggesting that their target was the antigen binding site of the T-cell receptor. This target was shown to be an epitope of the 90-kd heterodimeric T-cell receptor, which is composed of an α and β chain. The acidic α chain was shown to be disulphide linked to the more basic β chain and both chains were found to be glycoslated (Acuto *et al.*, 1983; Haskins *et al.*, 1984). Upon peptide analysis of the receptors from different T cells, it was found that each chain had portions that were variable and those that were constant, suggesting the presence of variable and constant domains, akin to those of immunoglobulin, in both the α and β chains (Kappler *et al.*, 1983; McIntyre and Allison, 1983).

Using the techniques of molecular biology, two independent groups successfully attacked the problem of defining the precise molecular nature of the T-cell receptor by cloning the message encoding the T-cell antigen receptor (Hedrick *et al.*, 1984; Yanagi *et al.*, 1984). This approach was based on a number of assumptions, the most fundamental of which was that the messages coding for the T-cell receptor would be T-cell-specific. As over 98% of the genes expressed in T and B cells are the same, a small number of T-cell-specific cDNA clones could be isolated, and then further screened for the properties that would be expected of T-cell receptor messages. By analogy to immunoglobulin, one could assume that the sequences coding for the T-cell receptor (TcR) would be organized as separate, noncontiguous sequences in the genome, which would somatically rearrange in T cells to produce a functional gene. Thus, hybridization to a probe complementary to a message encoding a T-cell receptor chain would reveal different genomic hybridization patterns in different T-cell lines, when compared to the germline pattern exhibited by the DNA of non-T cells. The T-cell receptor could also be expected to have a structure similar to that of immunoglobulin and that would be reflected in the significant homology to im-munoglobulin at the protein sequence level. Based on these assumptions, T-cell specific clones were isolated from man and mouse, using differential and sub-tractive cDNA hybridization and screened for TcR-encoding messages (Hedrick *et al.*, 1984; Yanagi *et al.*, 1984). The sequences encoding some of these cDNA clones were subsequently shown to rearrange in cells isolated from T-cell leuke-mias, as well as in both helper and cytotoxic T-cell lines (Toyonaga *et al.*, 1984). By comparison to immunoglobulin, the protein sequences deduced from these clones were shown to be composed of variable and constant regions linked by a joining segment (Siu *et al.*, 1984b). The general design of the T-cell recep-tor possesses a structure similar to that of the immunoglobulin light chains. A cartoon of the T-cell receptor, with the MHC class II product and an immu-noglobulin molecule for comparison, is presented in Fig. 1. Both the human and murine clones encoded the TcRβ chain since the deduced protein sequences matched that of the partial sequence of the N-terminal portion of that chain (Acuto *et al.*, 1984).

Subsequently, a number of different investigators isolated cDNAs encoding the α chain from mouse (Chien *et al.*, 1984a; Saito *et al.*, 1984) and man (Sim *et al.*, 1984; Yanagi *et al.*, 1985), using a variety of approaches. This chain is similar to the β chain, being composed of extracellular variable and constant region domains, a transmembrane region and a cytoplasmic tail. The deduced protein sequence of the α chain contains four and six potential sites for N-gly-

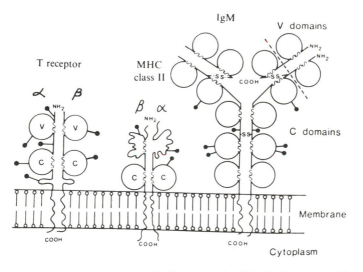

Figure 1. A pictorial representation of the T-cell receptor, class II major histocompatability complex (MHC) and immunoglobulin glycoproteins on the cell surface. These proteins have domain struc-tures that are either analogous to that of the constant region of immunoglobulin (C) or that of an immunoglobulin variable region (V). The balls on sticks represent potential N-glycosylation sites.

cosylation in the mouse and human respectively (Chien *et al.*, 1984a; Saito *et al.*, 1984; Sim *et al.*, 1984; Yanagi *et al.*, 1985), while the β chain has five and two potential sites for N-glycosylation in the mouse and human, respectively (Hedrick *et al.*, 1984; Yanagi *et al.*, 1984). The greater portion of both chains is extracellular and forms the recognition structure, while the transmembrane portions anchor the receptor in the cell membrane as part of a larger complex with the T3 proteins that are reputed to transduce the recognition signal (Acuto *et al.*, 1983; Kappler *et al.*, 1983; McIntyre and Allison, 1983; Haskins *et al.*, 1984). The extracellular portion of each chain is comprised of two domains. Each domain has an immunoglobulinlike structure of approximately 110 amino acids, with a central internal disulphide bond that spans approximately 65 amino acids and a number of conserved residues that are thought to be structurally important. The constant domain is essentially invariant, while the variable domains are different on different T cells (see Fig. 1). These variable domains are responsible for antigen recognition and contain the epitopes towards which the anticlonotypic monoclonal antibodies are directed.

Both the α and the β chains are members of the immunoglobulin gene superfamily, our knowledge of which has taken shape over the past few years. This family includes a number of molecules that are not involved in antigen recognition. (Strominger *et al.*, 1980; Mostov *et al.*, 1984; Clark *et al.*, 1985; Littman *et al.*, 1985; Maddon *et al.*, 1985; Nakauchi *et al.*, 1985; Sukhatme *et al.*, 1985), and even some which are not specific to the immune system (Williams and Gagnon, 1982; Chang *et al.*, 1985; Seki *et al.*, 1985), as well as the traditional immunological proteins, such as MHC products (Steinmetz *et al.*, 1981; Larhammar *et al.*, 1982; Malissen *et al.*, 1983), immunoglobulin, and the T-cell receptor. The members of this family share a common β-pleated sheet structure that consists of β-pleated sheets joined by an internal disulphide bond. The immunoglobulin and TcR genes are very closely related and this relationship is reflected not only in similarities in their sequences and protein structures, but also in their genomic structure, expression patterns, and regulation. A hypothetical evolutionary tree of the members of this family is presented in Fig. 2.

Both the α and β chains are expressed in helper and cytotoxic T-cell lines, but the α chain (Hedrick *et al.*, 1985; Toyonaga *et al.*, 1985), unlike the β chain, is also expressed in certain B-cell lines, in a fashion analogous to the expression of immunoglobulin in certain T-cell lines. Both chains contain a number of potential sites for N-linked glycoyslation and have similar structures with extracellular variable and constant domains, intracellular cytoplasmic tails, and a hydrophobic transmembrane region that contains a positively charged lysine. This unusual residue has been implicated in the assembly and export to the cell surface of the TcR/T3 complex, since cells that have lost the ability to produce a β chain cannot export T3 to the cell surface (Weiss and Stobo, 1984; Ohashi *et al.*, 1985), and the T3 proteins that span the cell membrane contain a negatively charged aspartic acid residue in their transmembrane region (van den

Elsen *et al.*, 1984). The assembly of a TcR/T3 complex is thought to neutralize these charges and allow the anchoring of the complex in the membrane.

The isolation of the sequences encoding both the α and β chains of the T-cell receptor opened up a number of avenues of research. The structure of

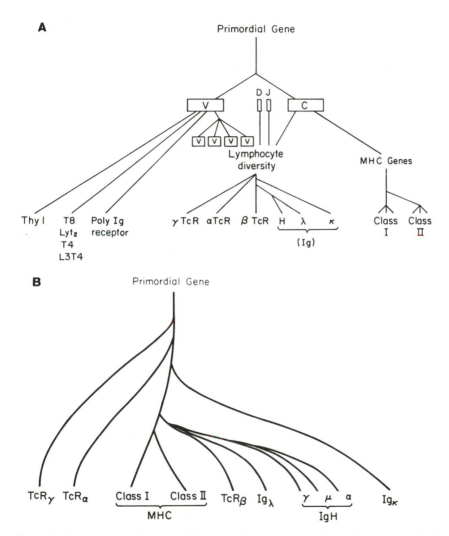

Figure 2. (A) A schematic diagram of the relationships between members of the immunoglobulin super gene family, based on the similarities in the domains analogous to those of immunoglobulin constant regions. (B) A diagram of a hypothetical evolutionary relationship between T-cell receptor, immunoglobulin, and MHC genes.

these molecules, and the role the different portions of the structure play in T-cell function, is now being examined. To begin with, the reconstitution of an active TcR by DNA transfer of the β chain indicated that the α-β heterodimer was indeed the T-cell antigen receptor (Ohashi *et al.*, 1985). This contention was further supported by the finding that transfection of α- and β-chain coding sequences conferred the specificity of the transferred receptor to the recipient cell (Dembic *et al.*, 1986; Saito *et al.*, 1987). This finding also supports the hypothesis that the α/β heterodimer alone is responsible for the specificity of antigen/ MHC recognition.

There has long been speculation as to whether T-cell recognition of antigen and MHC is mediated by a single receptor of dual receptors (see Fig. 3). The transfection results strongly support the hypothesis that there is only a single receptor for the recognition of antigen and MHC, and suggest that this receptor is composed of the α and β heterodimer.

The T-cell receptor genes have also been used to investigate the processes involved in the generation of diversity within the T-cell repertoire and to study the molecular basis of MHC restriction. In addition, events in the thymus, including the development of this restriction and the steps involved in the selection and maturation of the different classes of T cells, can now be probed. Another interesting area of investigation is the role the numerous invariant accessory

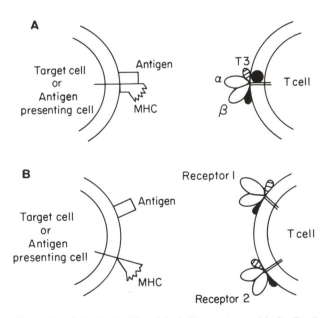

Figure 3. An illustration of the single (A) and dual (B) receptor models for T-cell recognition of MHC and antigen.

molecules play in T-cell recognition and function and their interaction with the T-cell receptor. These molecules include the T3 proteins, which seem to act as signal transducers for the TcR (Borst *et al.*, 1983; Meuer *et al.*, 1983), and the T4(L3T4) and T8(Lyt-2) molecules, which may bind invariant MHC determinants and aid in the stabilization of T-cell target interactions (Engleman *et al.*, 1981; Swain, 1981; Landegren *et al.*, 1982; Marrack *et al.*, 1983; Biddison *et al.*, 1984).

2. The Genomic Structure of the T-Cell Receptor Genes

The protein products of the immunoglobulin and T-cell receptor genes are composed of variable and constant domains. The constant region of immunoglobulin, and likely those of the TcR, mediates receptor-specific function, while the variable domains are responsible for the recognition of the myriad antigens by which the immune system is challenged. The diversity needed for this recognition is produced by the generation of a sequence encoding the variable domain, by the somatic recombination of different noncontiguous variable (V), joining (J), and diversity (D) gene segments. The genomic organizations of the sequences encoding the T-cell-receptor chains are similar to those of the immunoglobulin (Ig) gene clusters, with multiple copies of variable region gene segments located upstream of the segments encoding diversity, joining, and constant region sequences. However, the fine structures of the α, β, and immunoglobulin gene clusters differ greatly, in arrangement and number of gene segments, as well as in size and complexity of the clusters as a whole. A comparison of the T-cell receptor and immunoglobulin germline genomic organization is shown in Fig. 4. In T cells, the germline configurations of the T-cell receptor loci undergo somatic rearrangement to generate new and unique genomic structures, before a primary transcript can be made. This process (see Fig. 5) is similar to that involving the immunoglobulin loci during B-cell development. The choice of different V, D, and J segments is presumably random. An illustration of such a process is presented in Fig. 6 using the β-chain T-cell receptor genes as an example.

2.1. β-Chain Genes

Over the past few years the structure of the β-chain complex has been studied extensively. These genes, located on chromosome 7 in humans and 6 in mice (Caccia *et al.*, 1984a), are contained within a large stretch of DNA (Gascoigne *et al.*, 1984; Malissen *et al.*, 1984; Toyonaga *et al.*, 1984). The variable gene segments (V_β) are each composed of two exons. The shorter first exon is almost 50 nucleotides long and encodes leader sequences, while the 300-nucleotide second exon encodes the final five amino acids of the leader sequence and

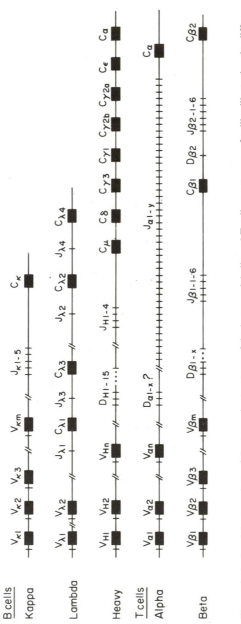

Figure 4. A pictorial comparison of the genomic structures of the immunoglobulin and T-cell receptor gene families illustrating the differences in the relative positions and numbers of constant, (C), joining (J), diversity (D), and variable (V) region gene segments.

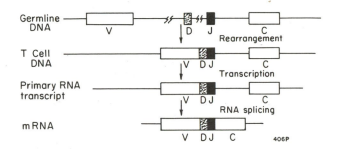

Figure 5. A pictorial representation of the steps involved in the generation of a transcript, which can be translated into protein, from germline T-cell-receptor β-chain DNA. Rearrangement first occurs between D_β and J_β segments and then a variable region is added to produce a fully rearranged gene. This gene is then transcribed to produce a primary transcript, out of which the sequences between the J_β and C_β segments are spliced, juxtaposing these regions in the final mRNA.

the majority of the variable domain (Siu *et al.*, 1984a). Both cysteines involved in the intradomain disulphide-linkage are encoded by the V_β gene segments.

While rearrangement studies with particular V_β genes indicate that many V_β are located upstream of the C_β genes, at least in the murine system, this is not always the case. One V_β is located 10 kb 3' to the 3'-most C_β gene in an inverted transcriptional orientation and has been shown to be used in a functional T cell (Malissen *et al.*, 1986). It is not known if there are other V_β 3' to the C_β segments in mice or if this arrangement is also present in the human genome. It has been estimated that there are approximately 20 β-chain variable (V_β) gene segments in mouse (Barth *et al.*, 1985; Behlke *et al.*, 1985) and over 100 in man (Concannon *et al.*, 1986; Kimura *et al.*, 1986). Unlike immunoglobulin heavy chain and κ variable region genes, which fall into large families of closely related genes, there is little cross-hybridization between members of either the murine or human V_β gene families (Barth *et al.*, 1985; Concannon *et al.*, 1986; Kimura *et al.*, 1986) and, while there are a higher number of human V_β segments in each family, the families are small, when compared to those of immunoglobulin. A possible explanation of this may be the more rapid divergence of these genes than has been found for immunoglobulin variable sequences.

Downstream of the V_β genes are two clonally related clusters, separated by just over 8 kb. Each cluster consists of a diversity (D_β) gene segment approximately 600 nucleotides 5' to a cluster of joining (J_β) region sequences, which are 2–5 kb 5' to a constant (C_β) region (see Fig. 7). The twelve nucleotide $D_{\beta 1}$ and fourteen nucleotide $D_{\beta 1}$ (Clark *et al.*, 1984; Siu *et al.*, 1984b; Toyonaga *et al.*, 1985) and almost 50 nucleotide J_β sequences (Gascoigne *et al.*, 1984; Malissen *et al.*, 1984; Toyonaga *et al.*, 1985) in man and mouse are homologous to each other and to the D and J segments of immunoglobulin genes. These areas of

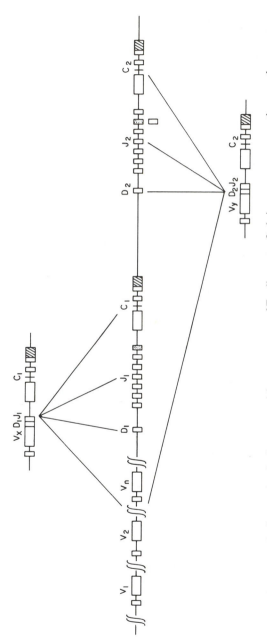

Figure 6. A schematic diagram showing how sequential rearrangements of T-cell-receptor β-chain genes can occur on the same chromosome and illustrating the potential for diversity that rearrangement provides.

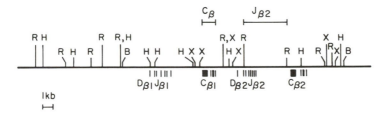

Figure 7. A germline, genomic restriction map of the human T-cell-receptor β-chain locus, showing variable (V_β), diversity (D_β), joining (J_β), and constant (C_α) region gene segments, using the restriction enzymes: BamHI (B), EcoRI (R), HindIII (H), and XbaI (X).

homology often contain the conserved residues that are thought to be structurally important for the variable domain. In both the murine and human genomes, the $J_{\beta1}$ cluster contains six functional J gene segments (Gascoigne *et al.*, 1984; Malissen *et al.*, 1984; Toyonaga *et al.*, 1985). The human $J_{\beta2}$ cluster has seven functional segments (Toyonaga *et al.*, 1985), while the murine genome has six functional $J_{\beta2}$ segments and one pseudogene between $J_{\beta2.5}$ and $J_{\beta2.6}$ (Gascoigne *et al.*, 1984; Malissen *et al.*, 1984). These structural and sequence homologies between the two clusters in man and mouse indicate that the clusters arose by a gene duplication event of a single cluster before the divergence of the two species.

In both mice and humans there are two C_β genes that are divided into 4 exons: the first two encode most of the extracellular constant domain; the third encodes a major part of the transmembrane region of the protein; and the last encodes the cytoplasmic coding sequences, as well as the 3' untranslated region of the β chain. The coding sequences of the two C_β genes, $C_{\beta1}$ and $C_{\beta2}$, are highly homologous, with four amino acid differences in mice and six in humans, concentrated at the carboxy-terminus of the protein, while the human introns' sequences bear little homology to each other or to the murine introns (Toyonaga *et al.*, 1985). These results suggest that, although there was extensive drifting between the clusters after the duplication event, there is strong pressure exerted for the retention of the coding sequences between the two C_β genes and between C_β genes of different mammalian species.

Recombinational signals, similar to those of immunoglobulin, are found flanking the V_β, D_β, and J_β genes (Clark *et al.*, 1984; Malissen *et al.*, 1984; Kavaler *et al.*, 1984; Siu *et al.*, 1984b). Immediately proximal to the coding segment is a highly conserved heptamer, followed by a nonconserved spacer, and then an A/T rich nonamer. The spacers can be either long or short with long spacers 3' to the V_β and D_β segments and short spacers 5' to the D_β and J_β segments (illustrated in Fig. 8).

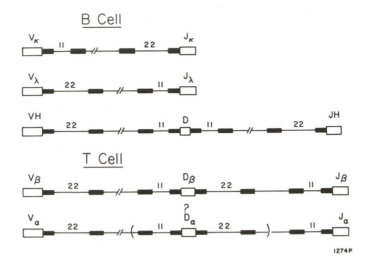

Figure 8. The distribution of long (22-nucleotide) and short (11-nucleotide) spacer sequences in the recombination signal sequences in immunoglobulin and T-cell-receptor genes. Since a sequence containing a short spacer is only permitted to recombine with one containing a long spacer, the distribution of spacers determines which segments can recombine. For example, rearrangement of immunoglobulin heavy chain genes is limited by the fact that a V segment can only join to a D segment, and not to a J, while in TcR$_\beta$, a V can join to either a D or to a J, thus increasing the potential for diversity.

2.2. α-Chain Genes

The α-chain genes are located on chromosome 14 in both man (Caccia *et al.*, 1985) and mouse (Dembic *et al.*, 1985). In man the region of chromosome 14 that contains these genes (14q11) is often rearranged in T-cell malignancies (Bernstein *et al.*, 1981; Hecht *et al.*, 1984; Williams *et al.*, 1984; Zech *et al.*, 1984), suggesting that in these cases the TcR α-chain locus may be involved in the activation of cellular protooncogenes, in a similar fashion to the activation of c-*myc* by translocation into the immunoglobulin loci (Caccia *et al.*, 1984b). Two cases of translocations involving the human c-*myc* locus and 14q11 have already been characterized (Shima *et al.*, 1986; Matieu-Mahul *et al.*, 1986). In the first (Shima *et al.*, 1986), the breakpoints occur between V$_\alpha$ sequences and 3′ to c-*myc* sequences, resulting in the translocation of C$_\alpha$, J$_\alpha$, and some V$_\alpha$ sequences to chromosome 8, 9–10 kb 3′ to the c-*myc* gene. In the second (Mathieu-Mahul *et al.*, 1986), the chromosome 8 breakpoint is less than 5 kb 3′ of the c-*myc* locus and the chromosome 14 breakpoint was localized to 14q11, with no known α sequences immediately proximal to the breakpoint, although this does not rule out the possibility of translocation into an uncharted region of the very large α-chain locus.

   The organization of the α-chain locus provides a striking contrast to those of β chain and immunoglobulin (see Fig. 9). A restriction map of the 40-kb stretch 5' to the C_α gene is illustrated in Fig. 10. There is a larger number (40–50) of variable region gene segments in both man (Yoshikai *et al.*, 1986) and mouse (Arden *et al.*, 1985), which can be divided into closely related cross-hybridizing families comprised of one to seven members (Yoshikai *et al.*, 1986), unlike the murine β-chain segments, which rarely cross-hybridize to each other. The V_α segments are also divided into two exons, the first coding for the signal peptide and the second for the last five residues of the signal peptide and for almost the first 100 amino acids of the variable domain (Hayday *et al.*, 1985; Winoto *et al.*, 1985; Yoshikai *et al.*, 1985). The coding sequences of the J_α regions are several codons longer than those of immunoglobulin and β chain (Arden *et al.*, 1985; Becker *et al.*, 1985; Hayday *et al.*, 1985; Winoto *et al.*, 1985; Yoshikai *et al.*, 1985, 1986) and are also more numerous and spread over a large distance (over 70 kb in mice and over 50 kb in man) (Hayday *et al.*, 1985; Winoto *et al.*, 1985; Yoshikai *et al.*, 1985). The J_α genes are separated on average by 1 kb (Hayday *et al.*, 1985), with the first located about 4 kb upstream of the single constant region (Hayday *et al.*, 1985; Winoto *et al.*, 1985).

   It would seem that there is only one α-chain constant region (C_α), which, like C_β, is composed of four exons. The first exon encodes a 9-amino-acid constant domain, which is short in comparison to those of β chain and immunoglobulin, and contains a cysteine bridge that spans only 49 amino acids, rather than the usual 65 amino acids. The second exon codes for the next 15 amino acids of the extracellular domain, which comprise part of the connecting peptide. The third exon encodes the transmembrane and cytoplasmic domains and the termination codon, while the fourth exon only contains 3' untranslated sequences, a feature not found in immunoglobulin or the T-cell receptor β-chain genes (Hayday *et al.*, 1985).

   Although no genomic sequences coding for D_α regions have been isolated as of yet, comparison of the sequences of two α-chain cDNA, isolated from HPB-ALL (Sim *et al.*, 1984) and Jurkat (Yanagi *et al.*, 1985), reveals a 15-nucleotide stretch in the former that is not present in the latter. It is possible that this stretch may code for a D_α segment, although the possibility of addition of these sequences by a mechanism analogous to the one responsible for N-region diversity in immunoglobulin cannot be ruled out.

   Recombination signals, homologous to those of immunoglobulin and β chain, are found flanking the α-chain segments, with long spacers 3' to the V_α genes and short ones 5' to the J_α gene segments (Hayday *et al.*, 1985; Winoto *et al.*, 1985; Yoshikai *et al.*, 1985). Since it has been shown that signals with short spacers only recombine with those with long spacers, it is possible for the V_α and J_α genes to join directly together without any need for a D segment, leaving the issue of D_α segments unresolved.

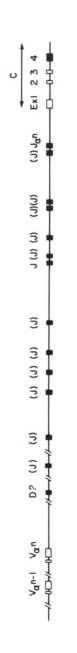

Figure 9. A pictorial representation of the T cell receptor α-chain locus in its germline state, containing variable (V_α), joining (J_α), and constant (C_α) region gene segments. The possible position of putative diversity (D_α) segments is also indicated.

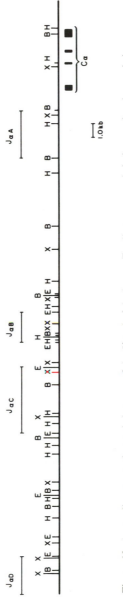

Figure 10. A germline, genomic restriction map of the 5′ end of the human T-cell-receptor α-chain locus using the restriction enzymes BamHI (B), EcoRI (E), HindIII (H), and XbaI (X).

3. Rearrangement of T-Cell Receptor Genes

The immune system of an animal must be able to deal with a wide variety of antigens. The required diversity of the recognition repertoire is mediated by T and B cells, each of which express a specificity different from that exhibited by another lymphocyte. The processes by which this diversity is generated have been well studied for immunoglobulin and, over the past two years, have been shown to be the same as those used for the generation of T-cell receptor diversity. The genes coding for these antigen-specific molecules are composed of separate, noncontiguous gene segments in the germline, which somatically rearrange to produce a functional gene, allowing for the generation of a wide variety of receptors by the combinatorial use of these segments (see Figs. 5, 6, and 7).

Both T-cell receptor and immunoglobulin gene rearrangement seem to be restricted to lymphocytes, and are fairly tissue specific, with no functional rearrangements of immunoglobulin genes in T cells or of T-cell receptor genes in B cells. A small percentage of T cells have partial rearrangements of heavy chain genes, but no light chain or complete heavy chain rearrangements (Cory *et al.*, 1980; Forster *et al.*, 1980; Kronenberg *et al.*, 1980; Kurosawa *et al.*, 1981; Kemp *et al.*, 1982; Kronenberg *et al.*, 1982; Nakanishi *et al.*, 1982; Zuniga *et al.*, 1982; Kraig *et al.*, 1983; Kronenberg *et al.*, 1983). A few B cells have what seem to be unproductive T-cell receptor gene rearrangements. An explanation for these inappropriate rearrangements is provided by the experiments of Yancopoulos *et al.* (1986), who introduced constructs containing D_β and J_β gene segments into B cells that rearrange their immunoglobulin genes. In these cells the D_β segment became joined to the J_β segment in a manner similar to that which would result in T cells, indicating that a similar, if not identical, enzyme system is instrumental in immunoglobulin and T-cell receptor gene rearrangement in B and T cells.

3.1. T-Cell Receptor Recombinational Signals

The rearrangement of T-cell receptor α, β, and γ gene segments is mediated by recognition signals similar to those of immunoglobulin (see Fig. 8). Immediately proximal to each of the coding V, D, and J segments is a highly conserved heptamer, followed by a nonconserved spacer, and then a conserved A/T rich nonamer. The spacers can be either long or short with their length roughly corresponding to either one (12 nucleotides) or two (23 nucleotides) turns of the DNA helix. It has been shown that a 12-nucleotide recognition sequence always is involved in recombination with a 23-nucleotide sequence, so that the distribution of these two spacers regulates which segments can be joined together.

An interesting example of the results of different spacer distributions is provided by a comparison between sequences flanking the D genes of T-cell receptor β chain and immunoglobulin heavy chain. In the heavy chain genes the signals 3' to the V_H genes and 5' to the J_H genes contain a long spacer, while there is a short spacer on either side of the D_H segment (Early et al., 1981). This arrangement of spacers makes the use of the D_H segment mandatory in the construction of a functional heavy chain gene, as there can be no joining of V_H to J_H segments, by the one turn/two turn rule. In the TcR β genes, however, the use of the $D_β$ segment is optional, as a result of an alternate arrangement of spacers, with long spacers 3' to the $V_β$ and $D_β$ segments, and short spacers 5' to the $D_β$ and $J_β$ segments (Malissen et al., 1984; Siu et al., 1984b) (see Fig. 8). This arrangement allows the $V_β$ segment to join either a $D_β$ or a $J_β$ segment. Examples of both cases of β-chain recombination have been found. The HPB-β2 message lacks a $D_β$ segment, while the YT35 (JUR-β1) messages contain a $D_β$ segment (Yoshikai et al., 1984). This arrangement of spacers also allows a $D_β$ segment to be joined to another $D_β$, instead of to a $J_β$ segment. The optional use of the $D_β$ segment, coupled with this possible use of multiple $D_β$ segments, can be seen as a means for the generation of added diversity of the T-cell repertoire. This diversity is increased by junctional flexibility, an imprecise joining process which effects the join at different nucleotides in the junction regions of $V_β$, $D_β$, and $J_β$ segments (Barth et al., 1985; Behlke et al., 1985; Concannon et al., 1986; Kimura et al., 1986; Kimura et al., 1987). N-region diversity, resulting from random addition of one to six nucleotides to either end of $D_β$ gene segments during recombination with $V_β$ and $J_β$ sequences, also occurs in T cells. Somatic hypermutation, which generates diversity in heavy and light chain immunoglobulin variable domains late in B-cell development by single base substitutions (Honjo, 1983; Tonegawa, 1983), has not been detected in T-cells as of yet.

3.2. β-Chain Rearrangement and Expression

Rearrangement of β-chain and immunoglobulin heavy chain genes seems to proceed in an orderly fashion with joining of D and J sequences preceding the joining of V sequences (Clark et al., 1984; Siu et al., 1984a; Siu et al., 1984b; Born et al., 1985). The most common rearrangement mechanism for immunoglobulin genes seems to be the formation of a stem–loop structure between recognition sequences, with the stem produced by basepairing between the heptamer and nonamer sequences and the loop comprised of the DNA between the segments being joined. Excision of the looped DNA between the joined segments leads to the joining of the segments and the deletion of the DNA sequences between them. This deletion of intervening sequences is seen in the majority of immunoglobulin heavy chain rearrangements (Alt et al., 1984) (see Fig. 11), but in the case of κ genes, there are rearrangement patterns that are incompatible with this mechanism (Hozumi and Tonegawa, 1976; Steinmetz et al., 1980; Selsing and Storb, 1981; Hochtl et al., 1982; Lewis et al., 1982; Van

1)

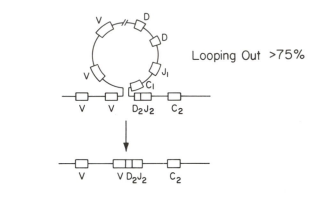

Looping Out >75%

2)

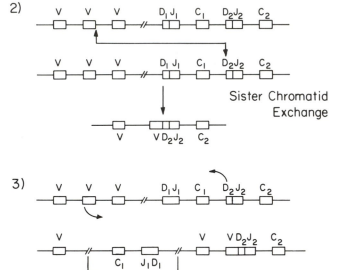

Sister Chromatid Exchange

3)

Inversion

Figure 11. Rearrangement mechanisms used in the generation of a complete T-cell-receptor gene. 1) Looping out/excision: DNA between the segments to be joined loops out and is then excised. This mechanism seems to be used in the majority of rearrangements. 2) Sister chromatid exchange: the arms of identical chromatids are exchanged during the rearrangement process. 3) Inversion: this process moves an inverted segment into a position beside a segment in the opposite orientation.

Ness *et al.*, 1982). The existence of these anomalous patterns led to the proposal of several other mechanisms, including unequal exchange between homologous sister chromatids (Selsing and Storb, 1981; Hochtl *et al.*, 1982; Van Ness *et al.*, 1982), inversion of DNA segments Hozumi and Tonegawa, 1976; Lewis *et al.*,

1982; Lewis *et al.*, 1984; Fedderson and Van Ness, 1985), and reintegration of excised sequences (Steinmetz *et al.*, 1980). Figure 11 illustrates the first three of these mechanisms.

In a study of a number of different T-cell receptor rearrangements, the majority of the rearrangement patterns could be explained by the looping/excision model, but over 30% of the patterns were not consistent with that model (Kronenberg *et al.*, 1984). In a number of cases intervening DNA sequences were retained, indicating any of the mechanisms described by the three alternate models could have been responsible, while in other cases, there were partial duplications of intervening sequences, suggesting unequal sister chromatid exchange (Duby *et al.*, 1985). In the case of the functional rearrangement involving the V_β 3' to $C_{\beta 2}$, inversion of the rearranged V_β and the 15 kb of DNA between the V_β and J_β was shown to occur (Malissen *et al.*, 1986). In this rearrangement, the D_β/J_β rearrangement seems to have occurred by unequal, but homologous, sister chromatid exchange. Thus, in addition to the looping/excision mechanism, all three alternate mechanisms seem to be used for the production of functional β-chain T-cell receptor genes, as illustrated in Fig. 11.

A large proportion of T cells have both β-chain alleles rearranged (Duby *et al.*, 1985), however, in all the cases examined to date there was only one functional V_β expressed in each clone examined, indicating that the mechanism of allelic exclusion is operating in these cells.

There are a number of complicated steps in the generation of a productive gene rearrangement and, at each of these steps, a change in the reading frame of the final product can generate a stop codon, rendering the rearrangement nonproductive. For example, only one of the three possible J_β reading frames is productive, so two-thirds of the $V_\beta-J_\beta$ joins put the J_β in an improper reading frame. This process is further complicated by the use of one or more D_β segments between the V_β and J_β. Another source of frameshift changes is the random addition of nucleotides at the V_β/D_β and D_β/J_β junctions that is employed to increase β chain diversity, but leads to a concomitant increase in the rate of nonproductive gene rearrangements. It is this high rate that provides an explanation of the low frequency of T cells that have only one β-chain rearrangement.

3.3. β Rearrangement in T-Cell Subsets

All helper and cytotoxic T cells examined so far have been shown to rearrange and express TcR α and β genes. The use of either of the two C_β regions does not correlate with either the class of T cell or the class of MHC molecule that is recognized, although the $C_{\beta 2}$ gene is used more often in the T-cell population as a whole. Examination of β-chain expression in murine suppressor cell hybridomas revealed that a large proportion of these hybrids had deleted the whole β gene cluster from both their chromosomes (Kronenberg *et al.*, 1984;

Hedrick *et al.*, 1985), raising a number of questions as to the role β chain plays in the function of this T-cell class.

3.4. β-Chain Transcripts

There are two sizes of β-chain mRNA: a full-length 1.3-kb transcript and a 1.0-kb transcript, which contains no V_β sequences (Yoshikai *et al.*, 1984). The shorter transcript can be derived from two different genomic arrangements. In a number of cases, the short transcript is produced from a rearrangement that joins a D_β segment to a J_β, without a V_β gene (the first step in the rearrangement process) and consists of D_β, J_β, and C_β sequences. The other type of short message is derived from an unrearranged locus and consists of only J_β and C_β sequences, which are juxtaposed by RNA processing of the primary transcript.

3.5. α-Chain Rearrangement and Transcription

Expression of α chain has been demonstrated in both helper and cytotoxic T-cell lines, as well as in T-cell tumors. In mature T cells the level of α and β transcription are similar, although in the thymus β-chain levels are higher than those of α in immature populations (Chien *et al.*, 1984a; Raulet *et al.*, 1985; Haars *et al.*, 1986). Two sizes of α-chain transcripts have been described: a full length 1.7-kb message and a shorter 1.4-kb transcript. By analogy to the β-chain transcripts, the shorter one could be the product of incompletely rearranged or unrearranged sequences. This hypothesis is strengthened by the finding of a number of α-chain cDNA clones with only J_α and C_α sequences (Yoshikai *et al.*, 1986).

The study of α-chain gene rearrangement has been hampered by the sheer size of the DNA over which the J_α genes are spread, since there is no one probe that can encompass the sites of productive gene rearrangement. Rearrangement can be shown by the use of J_α-region probes (Sangster *et al.*, 1986) or V_α segments, but this method is very laborious. A more complete map of the α-chain locus detailing where segments are in relation to each other will be of great help in this task, and a more complete analysis will have to await the cloning of all the J-region segments.

4. The Generation of T-Cell Receptor Diversity

In the course of an organism's lifetime, its immune system must deal with a vast number of foreign antigens. The problem of generating the necessary diversity of antigen receptors, without overburdening the genome with too many different sequences, has been dealt with by both B and T cells in a number of interesting ways.

In B cells, three basic mechanisms provide a diversified and adaptive response to foreign antigen: germline diversity, combinatorial diversity, and somatic mutation. Germline diversity is the presence of a large number of variable, joining, and diversity segments from which different variable domains can be constructed. Combinatorial diversity arises from the random rearrangement of these different germline sequences, while somatic mutation produces base-pair changes both during the formation of the variable domain and during the maturation of the immune response, to provide a receptor with higher affinity for the target antigen, a process known as somatic hypermutation.

These mechanisms, with the possible exception of somatic hypermutation (Ikuta *et al.*, 1985), are also responsible for T-cell receptor diversity, but the extent to which they contribute to the total diversity varies between the two chains.

5. The T-Cell Receptor Repertoire

Estimates of the size and diversity of the T-cell receptor V-region repertoires have been obtained by the use of two methods; genomic Southern analysis using V-region probes and statistical analysis of frequency of V-region isolation from cDNAs. The former technique probes Southern blots of enzyme-restricted germline genomic DNA with noncrosshybridizing V-region segments, and then tabulates the number of different bands obtained. Use of different enzymes, to generate different patterns, and careful analysis of band sizes and numbers can be used to determine the number of members in a given crosshybridizing family and the size of the total repertoire. Using the latter technique, quantitative estimates of the number of V regions are derived from statistical analysis of the frequency of isolation of a given V region in cDNAs isolated from various sources. The reliability or significance of the estimates is dependent on two major factors. First, the cDNAs must be randomly sampled with no bias toward cDNAs isolated from a particular population. Second, the cDNAs must be derived from genes that are utilized roughly equally. The accuracy of the estimates is highly dependent on the algorithms used to analyse the tabulated frequencies of isolation, and ideally requires a large sample size in which the number of unique observations is significantly less than the total number.

Using statistical analysis, estimates for the human V_α and V_β repertoires have been tabulated and are shown in Table I. Calculations have been performed on data obtained from single individuals (HAVP, HAVT, HBVP, and HBVT) as well as on all of the data available. In the aggregate data analysis cDNAs with identical V, D, J, and C regions are counted only once. These estimates are only approximate, since the effects of somatic hypermutation and the presence of polymorphism in the highly heterogeneous human population cannot be assessed.

Table I. Statistical Estimates of V Gene Segment Repertoires

TcR α chain:	Source[a]	Number of cDNA[b] containing V_α	Number of unique V_α	Most probable number of B_α (probability)	95% upper bound of V_α repertoire
	HAVP	18	13	24 (0.28)	60
	HAVT	10	8	19 (0.37)	112
	all[c]	31	23	47 (0.21)	92

TcR β chain:	Source[a]	Number of cDNA[b] containing V_β	Number of unique V_β	Most probable number of V_β (probability)	95% upper bound of V_β repertoire
	HAVP	10	9	42 (0.43)	881
	PL[d]	26	18	32 (0.24)	61
	PH	15	12	30 (0.31)	110
	HBVT	11	8	15 (0.36)	54
	all[c]	76	46	68 (0.15)	90

[a]Abbreviations: see Figures 13 and 14.
[b]Each cDNA is made up of a unique combination of V, D, J, C, and N region sequences.
[c]All cDNA/genomic data available.
[d]This data was collected from libraries derived from more than one individual.

The murine and human TcR V_α repertoires both contain at least 50 members which can be grouped into over a dozen families (Arden *et al.*, 1985; Becker *et al.*, 1985; Yoshikai *et al.*, 1986). There are at least 70 human TcR V_β genes in over 20 families (Concannon *et al.*, 1986; Kimura *et al.*, 1986), while the murine TcR V_β repertoire is considerably smaller, with 20–25 V_β genes (Barth *et al.*, 1985; Behlke *et al.*, 1985). There may be an even larger number of human V_β genes (see HBVP in Table I), since the data collected in Table II and presented in Table VIa, indicate that a subset of V_β families is used more frequently than the rest.

The hybridization patterns of human germline DNA probed with selected TcR α and β chain cDNAs are shown in Figs. 12a,b. Estimates of the V region repertoires based on the numbers of bands that do not contain constant region sequences, are consistent with those arrived at by statistical analysis. Crosshybridization studies agree with the empirical rule that the crosshybridization threshold is roughly equivalent to 75% homology at the nucleotide sequence level (see HBVP25 and HBUP50 in Fig. 12).

In the analysis of the V_α sequence data, the cDNA and genomic clones (from which the introns have been removed) have been aligned with each other. To maximize areas of homologous residues, spaces have been added to the nucleotide sequences and in the equivalent place in the corresponding deduced amino acid sequence. The alignments of the deduced amino acid sequences are shown in Fig. 13. The frequency of the most common residue (consensus residue) at a given position (the vertical columns in Fig. 13) was calculated as a

Table II. Gene Segment Composition of TcR B-Chain cDNAs

Entry	V_β	D_β	J_β	C_β	Clone	Comments
1	1.1	1.1/2.1	2.3	2	PL5.6	
2	1.1	2.1	2.3	2	PL6.4	
3	1.1	1.1/2.1	2.1	2	PL6.1	
4	1.1	2.1	2.1	2	PL5.2	
5	1.2	1.1	1.5	1	HBVT96	
6	1.2	2.1	2.2	2	HBVT73	
7	2.1	—	2.1	2	PL2.13	
8	2.1	2.1	2.1	2	PL6.21, MOLT4	
9	2.2	2.1	2.3	2	MT11	
10	2.3	—	1.1	NR	PH34	
11	2.3	1.1	1.3	NR	PH7	
12	3.1	1.1	2.5	2	PL4.4	
13	3.1	1.1	1.5	1	PL4.22	
14	3.1	—	1.6	1	PL3.10	
15	3.1	2.1	2.5	2	HBVT22	
16	3.2	1.1	2.1	2	DT259	
17	3.3	2.1	2.7	2	HBVP55	Different N region sequences.
18	3.3	2.1	2.7	NR	PH21	
19	3.3	—	2.3	2	PL8.1	
20	4.1	2.1	2.3	2	PL2.14	1 nucleotide difference in V.
21	4.1	—	1.1	1	PL5.7	
22	4.1	2.1	2.5	NR	2G2	
23	4.2	1.1	1.1	1	DT110	
24	4.3	1.1	2.3	2	HBVP48	
25	5.1	1.1	1.1	1	PL7.16	Different N region sequences.
26	5.1	1.1	1.1	1	PL4.16	
27	5.1	2.1	2.3	2	HPVP51	
28	5.2	—	2.2	2	PL2.5	
29	5.3	2.1	2.5	2	12A1, HPB-ALL, HPBβ2	
30	5.4	2.1	2.5	NR	PH24	
31	6.1	1.1	2.1	2	PL4.14	
32	6.1	2.1	2.5	2	HPB-MLT, 4D1	
33	6.1	1.1	1.1	1	HBVPO4	
34	6.1	2.1	2.3	2	HBVP50	
35	6.2	1.1	1.5	1	PL5.10	
36	6.3	1.1	1.5	1	ATL122	
37	6.3	—	2.7	NR	PH5	
38	6.3	2.1	2.1	2	HBVT23	
39	6.3	2.1	2.7	2	HBVT10, HBVT41, HBVT65	
40	6.4	2.1	2.5	2	HBVP25	
41	6.5	—	2.1	2	HBVT16	
42	6.5	—	2.7	2	HBVT11	
43	6.6	1.1	1.5	1	HBVT45	
44	6.7	2.1	2.7	NR	PH16	
45	6.7	2.1/1.1	2.6	NR	PH79	
46	6.8	—	1.3	NR	PH22	
47	6.9	1.1	1.5	NR	L17	
48	7.1	1.1	1.1	1	PL4.9	
49	7.2	1.1/2.1	2.3	2	PL4.19	

(continued)

Table II. *(Continued)*

Entry	V_β	D_β	J_β	C_β	Clone	Comments
50	8.1	1.1	1.2	1	YT35	
51	8.1	1.1	1.1	NR	PH11	
52	8.2	1.1	1.4	1	PL3.3	
53	8.2	—	—	-	HBV32SP	
54	8.3	NR	NR	NR	HBVP41	
55	8.4	2.1	2.6	NR	PH8	
56	9.1	2.1	2.6	2	PL2.6	
57	9.1	—	2.3	-	CEM2	
58	10.1	1.1	2.1	2	PL3.9	
59	10.2	1.1	1.5	1	ATL121	
60	11.1	—	2.2	2	PL3.12	
61	11.2	—	2.3	NR	PH15	
62	12.1	—	2.5	2	PL4.2	
63	12.2	1.1	1.3	1	HBVP54	
64	12.2	—	2.5	NR	PH27	
65	12.3	—	1.1	1	PL4.24	
66	12.3	1.1	1.2	1	HBVP34	
67	12.4	1.1	1.5	1	PL5.3	
68	12.4	NA	NA	NA	CEM1	
69	15.1	1.1	1.5	1	ATL21	
70	15.1	2.1	2.7	NR	PH32	
71	16.1	1.1	1.2	1	HBVP42	
72	17.1	—	2.1	2	HBVTO2	
73	18.1	—	2.1	2	HBVT56	
74	18.2	1.1	1.4	NR	PH29	
75	18.2	2.1	2.1	NR	PH26	
76	19.1	1.1	1.6	1	HBVT72	
77	20.1	—	1.2	NR	HUT	
78	P	P	2.4	2	HBVP22, HBVP68	
79	P	2.1	2.1	2	HBVP15, HBVP37	Different N region sequences.
80	P	2.1	2.1	2	HBVP31	
81	—	—	2.5	2	HBVP58, HBVP63	
82	P	P	2.1	2	PL3.1	

*a*Abbreviations: NA, not available; NR, not reported; —, unassigned; P, partial rearrangement products (DJC, JC).

percentage of the total number of residues. The consensus sequence is displayed as a function of frequency from 10% to 100% above the alignment. Pairwise comparison of V_α nucleotide sequences, excluding the leader residues, allows the identification of sequences that share over 75% homology and their subsequent grouping into families whose members would crosshybridize (Yoshikai *et al.*, 1986). The family assignments given in Fig. 13 and 14 maintain the established nomenclature (Yoshikai *et al.*, 1986; Kimura *et al.*, 1987) in which unique members of the same family share the first digit and differ in the second. For example, $V_\alpha 1.1$ and $V_\alpha 1.2$ are two crosshybridizing members of the $V_\alpha 1$ family.

A similar analysis of the human V_β repertoire has been performed. The alignments of the deduced amino acid sequences of unique V_β chains are shown in Fig. 14, and a more complete compilation of the usage of different V, D, J, and C gene segments, derived from the current literature, is presented in Table III. The nomenclature used is derived from that of Kimura *et al.* (1986) and Concannon *et al.* (1986). Since the publication of their papers, new V_β sequences have been isolated, new families have been defined, and old ones deleted as the result of amalgamation of families (Kimura *et al.*, 1987). For example, members of the $V_\beta 13$ and $V_\beta 14$ families have merged respectively with the $V_\beta 12$ and $V_\beta 3$ families. The convention adopted, to minimize confusion, was to merge members into the lower numbered family and to leave the vacated family empty. The consensus sequence analysis is presented above the alignments. Certain sequences (designated by small capitals in Fig. 14) have been omitted from the consensus calculations, but have been assigned unique family designations for completeness. While it is possible that those sequences that differ only in their leader portions ($V_\beta 1.2$, 4.1, 6.6, 10.2, 11.2, 18.2) originated from unique germline sources, they may only be polymorphic variants of the same genes. $V_\beta 6.2$ and $V_\beta 8.3$ cannot be uniquely assigned as a result of insufficient sequence data.

As can be seen in Figs. 13 and 14, unlike Ig genes that have the well-defined hypervariable and framework regions (Wu and Kabat, 1970), the α- and β-chain genes show more diffuse regions of conservation and variability. However, highly conserved framework residues are found in both TcR and Ig genes.

The J_α sequences obtained from cDNA sequencing are presented in Fig. 15. Each unique J_α, denoted by large capital letters in Fig. 15, was used in consensus calculations the results of which are shown above the alignments, but all reported J_α have been included for comparison. A germline V_α sequence is included in the figure to indicate the approximate location of the J_α/V_α boundary, and to help illustrate the extent of functional diversity. Since only six human J_α have been mapped with respect to the C_α gene ($J_\alpha A$–$J_\alpha F$ in Fig. 15) (Yoshikai *et al.*, 1985), each unique J_α has been assigned a letter upon sequencing and comparative analysis to await ordering in the germline. Table IVb summarizes the subgroups of J_α genes and their frequency of use in the cDNAs sequenced to date. Estimation of the human J_α repertoire, by statistical analysis, predicts that there are 30–60 human J_α with a 95% upper bound of 150–200 different genes. A similar analysis of human J_β gene usage is shown in Table IV. All 13 known germline segments are utilized and their frequency of usage indicates that there are no other J_β segments in the genome.

While two β-chain diversity gene segments have been isolated from the human genome (Yoshikai *et al.*, 1985), it is difficult to identify them by cDNA analysis, as a result of the combined effects of junctional flexibility, N-region diversity, the brevity of D_β segments, and the presence of multiple D_β translational reading frames. Thus, it is unlikely that the existence of D_α segments will be deduced from analysis of α-chain cDNA sequences.

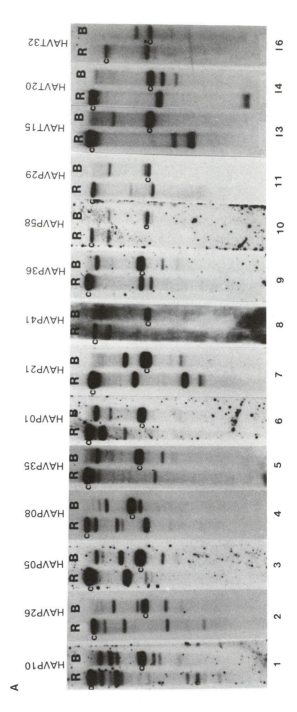

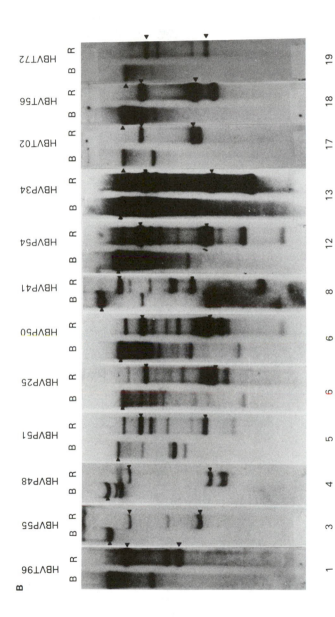
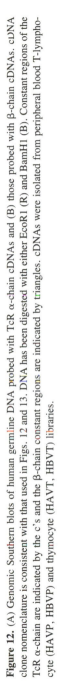

Figure 12. (A) Genomic Southern blots of human germline DNA probed with TcR α-chain cDNAs and (B) those probed with β-chain cDNAs. cDNA clone nomenclature is consistent with that used in Figs. 12 and 13. DNA has been digested with either EcoR1 (R) and BamH1 (B). Constant regions of the TcR α-chain are indicated by the c's and the β-chain constant regions are indicated by triangles. cDNAs were isolated from peripheral blood T-lymphocyte (HAVP, HBVP) and thymocyte (HAVT, HBVT) libraries.

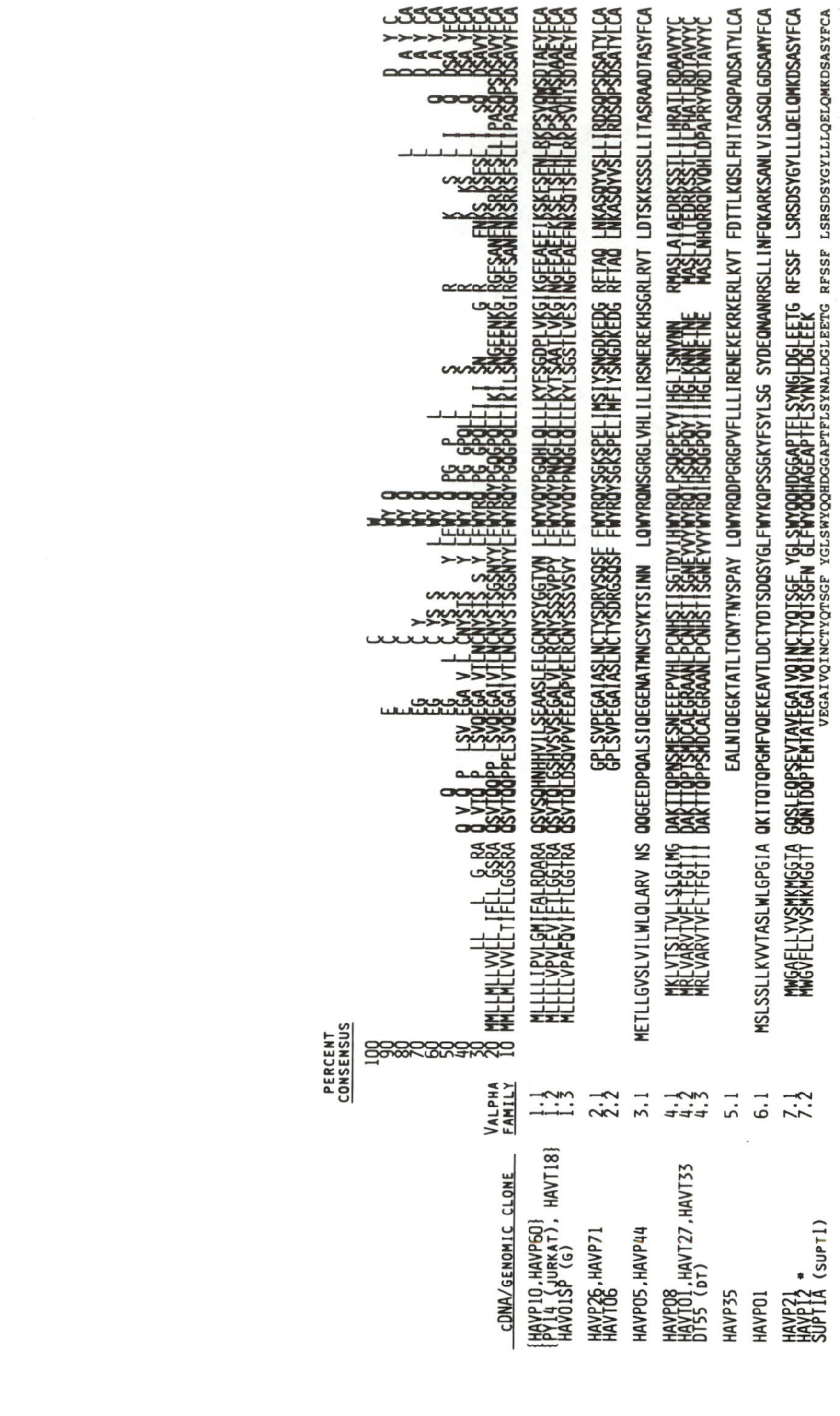

PERCENT
CONSENSUS

100
90
80
70
60
50
40
30
20
10

CDNA/GENOMIC CLONE	VALPHA FAMILY						
[HAVP10,HAVP50]	1.1						
[PV14 (JURKAT), HAVT18]	1.2						
HAVO1Sβ (G)	1.3						
HAVP26,HAVP71	2.1						
HAVT06	2.2						
HAVP05,HAVP44	3.1						
HAVP08	4.1						
HAVT01,HAVT27,HAVT33	4.2						
DT55 (DT)	4.3						
HAVP35	5.1						
HAVP01	6.1						
HAVP21 *	7.1						
SUPT1Ā (SUPT1)	7.2						

		leader	Vα
{HAVP41,HAVP171,HAVP49 / HAVP50,HAVP172,HAVP24}	8.1 / 8.2	MTSIRAVFIFLMLQLDLVNG / MAGIRALFMYLMLQLDWSRG	ENVEQHPSTLSVQEGDSAVIKCTYSDSASNYFP / ESVGQHPSTLSVQEGQNSTINCAYSNSASDYFI ... MXYQELGKRPQFIIDIRSNVGEKKDQQRLRVT LNKTAXHFSLHITETQERDSSAVYFCA / MXYBESGKBPQFIIDIRSNMDKRQGQRVTVL RSNMDKRQGQRVTVL IAATQPBDSSAVYFCA
HAVP36	9.1	MKPTLISVLV1IFILRGTRA	QRVTQPEKLLSVFKGAPVELKCNYSYSGSPE LFMYVQYSRQRLQLLLRHISRES IKGFTADLNKGETSFHLKKPFAQEEDSAMYYCA
HAVP58	10.1	MVLKFSVS1LMIQLAMVS T	QLLEQSPQFLSIQEGENLTVYCNSSSVFSS LQWYRQEPGEGPVLLVTVVTGGEVKKLKRLTFQ FGDARKDSSLH1TAAQPGDTGHYLCA
{HAVP02,HAVP28 / HAVP29,HAVP32}	11.1	MALQSTLGAVWLGLLLNSLWKVAES	KDQVFQPSTVASSEGAVVE1FCNHSVS NAYNFFWYLHFPGCAPRLLVKGSKPSQQGR YNMTYERFSSSLI1LQVREADAAVYYCA
PGA (HPBMLT)	12.1	IFASLLRAVIASICVVSSMA	QKVTQAQTEISVVEKEDVTLDCVYETRDTTYYLFWYKQPPSGELVFLIRRNSFDEQNEISGRYSWNFQKSTSSNFT1TASQVVDSAVYFCA
HAVT15	13.1	PGALLGLLSAQVCCVRG	IQVEQSPPDLILQEGANSTLRCNFSDSVNN LQWFHQNPWGQLINLFYIPSGTKQN GRLSATTVATERYSL LYISSSQTTDSGVYFCA
HAVT20	14.1	MACPGFLWALVISTCLEFSMA	QTVTQSQPEMSVQEAETVTLSCTYDTSESDYYLFWYKQPPSRQMILVIRQEAYKQQNATENRFSVNFQKAAKSFSLKISDSQLGDAAMYFCA
HAVT31	15.1	MRQVARVIVFLTL SMSRG	EDVEQSL FLSVREGDSSVINCTYTDSSTY LYWYKQEPGAGLQLLTYIFSWNDMKQDQRLTVL LNKKDKHLSLRIADTQTGDSAIYFCA
{HAVT32,HAVT35} **	16.1	MASAPISMLAMLFTLSGLRA	QSVAQRKIRSTLLKGIL*TVKCTYSVSGNPY LFWYVQYPNRGLQFLLKYITGDNLVKGSYGFEAEFNKSQTSFHLKKPSALVSDSALYFCA

Figure 13. Deduced amino acid sequences of human TcR V_α segments. Spaces at equivalent positions in the nucleotide and deduced amino acid sequences of a given cDNA clone were added to maximize regions of homology. HAVP12(*) has a deletion of the 3' V and the entire J gene segment so that C_α is found directly attached to the partial V gene. **An apparent single nucleotide deletion has led to a translational reading frameshift (●). A genomic V_α sequence (G) has been included along with cDNA sequences for comparison.

CDNA/GENOMIC CLONE	V BETA FAMILY	Sequence
PL5.6	1.1	MWQDLLAWGWCLLGAGPV DSGVTQTPKHLITATGQRVTLRCSPRSGD LSVYWY QQSLDQGLQFLIQYYNGEE RAKGNILERFSAQQFPDLHSELNLSSLELGDSALYFCASS
HBVT73	1.2	MGFRLLCCVAFCLLGAGPV DSGVTQTPKHLITATGQRVTLRCSPRSGD LSVYWY QQSLDQGLQFLIQYYNGEE RAKGNILERFSAQQFPDLHSELNLSSLEGDSAL/FCASG
PL2.13	2.1	MLLLLLLGPAGSGL GAVVSQHPSWV ICKSGTSVK IEERSLDFQATIHFMY RQFPKQSLMLMAISNEGSKATYEQGVEQKFLINHASLTISTLTVTSAHPEDSSFYICSAS
M.11 (G)	2.2	MLLLLLLLGPSLAGSGL GAVVSQHPSWV ICKSGTSVK IEERSLDFQATIHFMY RQFRKXSLMLEAISNEGSKATYEQGVEQKFLINHASLTISTLTVTSAHPEDSSFYICSAK
PH34	2.3	MLLLLLGPASSGL GAVVSQHPSRV ICKSGTSVK IEERSLDFQATIHFMY RQFRKXSLMLEGAISNEGSKATYEQGVEQKFLINHASLTISTLTVISAHPEDSSFYICSA
HBVT22	3.1	GIRLLCRVAFCFLAVGLV DVKVTQSSRYLVRTGEKVFLEVCQDMDH EMFEKY RQDPGLGLRLIYFSYDVK MKEKGDIPEGYSVSREKKERFSLILESASINQTSMYLCASS
DT25g (DT)	3.2	AVGLV DVKVTQSSRYLVKRTGEKVFLEVCQDMDH EMFEKY RQDPGLGLRLIYFSYDVK VTDKGDVPEGYSVSRKKEKRNFPLILESSPNQTSLYFCASS
HBVP55	3.3	VTCSQWNH EMKSWY RQDPGLGLRQIYFSYMNVE
PL2.14	4.1	MSDPKLVGPHEYL SGLGSVF SAVISQKPSRDICQRGTSLT IQCQVDSQV TMFEKY RQDPGQSLTLIATANQGSEATYESGFVIDKPISRPNLTFSTLTVSNMSPEDSSTYLCSSVR
DT110 (DT)	4.2	SAVISQKPSRDICQRGTSLT IQCQVDSQV TMFEKY RQDPGQSLTLIATANQGSEATYESGFVIDKPISRPNLTFSTLTVSNMSPEDSSIYLCSSR
HBVP48	4.3	SGLGSVF TMFFKY RQDPGSSLTLIATANQGSEATYESGFVIDKPISRPNLTFSTLTVSNMSPEDSSIYLCSAG
HBVP51	5.1	MGSRLLCWLLCLLGAGPV KAGVTQTPRYLIKTRGQVTLSCSPISGH RSVSWY QOTPGQGLQFLFEYFSETQ RNKGNEFPQRESGQFSNSRSEMNVSTLELGDSALYLCASS
PL2.5 (HPBMLT)	5.2	AALHL IKTRGQHCTLRCSPISGH QGVTLSCSPISGH RSVSWY QGVLGQGLQFLFEYFSETQ RQRGNFPDRFSAROFPNYSGELNVAALLGDSALYLCASS
PH24	5.3	RRHCWVLLCLLGAGPV RAGVTQTPRHLIKTRGQNVTLGCSPISGH RSVSNY QOTLGQGLQFLFEYFSETQ RNKGNFLDRFSARQFPNSGELNVSTLELGDSALYLCASA
HBVP50	5.4	MGTRLLCWAALCLLGADHT GAGVSQTPSNKVTEKGKYVELRCDPISGH TALYMY RQSLGQGPEFLIYFQGTG AADDSGLPNDRFFAVRPEGSVSTLKIQRTERGDSAVYLCASS
PL5.1	6.1	MGTRLLCWVLGFLGTDH GAGVSQSPRYKVAKRGQDVALRCDPISGHV SLFMY RQALGQGPEFLTYFQNEAQ LDKSGLPSDRFFAERPEGSVSTLKIQRTQEDSAVYLCASS
HBVP25	6.2	GAGVSQSPRYKVAKRGQDVALRCDPISGHV SLFMY RQALGQGPEFLTYFQNEAQ LDKSGLPSDRFSAEPEGSVSTLKIQRTQEDSAVYLCASS
HBVT116	6.3	MALCLLGADHA DTGVSQNPRHKITKRGQNVTFRCDPISEH NRLYWY RQALGQGPEFLTYFQNEAQ QDKSGLPNDRFSAERPEGSVSTLEIQRTEQGDSAMYRCAST
PH16	6.4	VSTDHT GAGVSQSPSNKVTEKGKDVERCDPISGH TALYWY RQSLGQGPEFLIYFQGNS APDKSGLPSDRFSAERTGGSVSTLTIQRTQQEDSAVYLCASS
L17 (L17)	6.5	MGTSLLCWMALCCLLGADDE ISGVSQNPRHKITKRGQNVTFRCDPSEH NRLYWY RQALGQGPEFLTYFQNEAQ LEKSRLLSDRFSAERPKGSFSTLEIQRTEGGDSAMYLCASS

(amino acid sequence alignment of human T-cell receptor V beta region gene segments; PERCENT CONSENSUS scale 100–10 shown above the alignment with consensus residues)

PL4.9	7.1		HLVMGMTNKKSLKCEQHMGH RAMYWY KQKAKKPPELMFVYSYEK LSINE SVPSRFSPECPNRSLLNLHLHALQPEDSALYLCASS
PL4.19	7.2		SSGFSPECPNSSHLFLHLHTLQPEDSALYLCASS
YT35 (JURKAT)	8.1	MDSWTFCCVSLCILVAKHT	DAGVIQSPRHEVTEMGQEVTLRCKPISGH NSLFWY RQTMMRGLELLIYFNNNV PIDDSGMPEDRFSAKMPNASFSTLKIQPSEPRDSAVYFCASS
PL3.3	8.2		EVTEMGQEVTLRCKPISGH DYLFWY RQTMMRGLELLIYFNNNV PIDDSGMPEDRFSAKMPNASFSTLKIQPSEPRDSAVYFCASS
HBVP41	8.3	MGSWTLCCVSLCILVAKHT	DAGVIQSPRHEVTEMGQEVTLRCKPISGH DYLFWY RQTMMRGLELLIYFNNNV PIDDSGMPEDRFSAKMPNASF
PH8	8.4	MGSWTLCCVSLCILVAKHT	DAGVIQSPRHEVTEMGQEVTLRCKPISGH DYLFWY RQTMMRGLELLIYFNNNV PIDDSGMPEDRFSAKMPNASFSTLRIQPSEPRDSAVYFCASS
PL2.6	9.1		TELGNDKSIKCEQNLGH DTMYWY KQDSKKFLKIMFSYNNKE LIINE TVPNRFSPKSPDKAHLNLHINSLELGDSAVYFCASS
PL3.9	10.1	VGSDFSCCVAISFWGARLT	DTKVTQRPRLLVKASEQKAKMDCVPIKAH SYVYWY RKKLEEELKFLVYFQNEE LIQKAEIINERFLAQCSKNSSCTLEIQSTESGDTALYFCASS
ATL 121 (G)	10.2	MCLRLLCCVAISFWGA ST	DTKVTQRPRLLVKASEQKAKMDCVPIKAH SYVYWY RKKLEEELKFLVYFQNEE LIQKAEIINERFLAQCSKNSSCTLEIQSTESGDTALYFCASS
PL3.12	11.1	MTIRLLCYMGFYFLGAGLM	EADIYQTPRYLVIGTGKKITLECSQTMGH DKMYWY QQDPGMELHLIHYSYGVN STEKGDLSSESTVSRIRTEHFPLTLESARPSHTSQYLCASM
PH15	11.2	MAQFGLLFSGAGLM	EADIYQTPRYLVIGTGKKITLECSQTMGH DKMYWY QQDPGMELHLIHYSYGVN STEKGDLSSESTVSRIRTEHFPLTLESARPSHTSQYLCASS
PL4.2	12.1		GTPVTLRCHQTENH RYMYWY RQDPGHGLRLIHYSYGVKD TDKGEVSDGYSVSRSKTEDFLLTLESLPAPETSVYFCATR
HBVP54	12.2		LRCHQTENH RYMYWY RQDPGHGLRLIHYSYGVKD TDKGEVSDGYSVSRSKTEDFLLTLESATSSQTSVYFCANG
HBVP34	12.3	MSIGLLCCAALSLLWAGPV	NAGVTQTPKFQVLKTGQSMTLQCAQDMNH EYMSWY RQDPGMGLRLIHYSVGAG ITDQGEVPNGYNVSRSTTEDFPLRLLSAAPSQTSVYFCASS
CEM.1 (G)	12.4	MSLGLLCCGAFSLLWAGPV	NAGVTQTPKFRVLKTGQSMTLLCAQDMNH EYMYWY RQDPGM*LRLIHYSVGEG TTAKGEVPDGYNVSRLKKQNFLLGLESAAPSQTSVYFCASS
ATL 21 (G)	15.1	MASLLFFCGAFYLLGTGSM	DADVTQTPRNRITKTGKRIMLECSQTKGH DRMYWY RQDPGLGLRLIYYSFDVKD INKGEISDGYSVSRQAQAKFSLSLESAIPNQTALYFCATS
HBVP42	16.1		QTVTLRCDPISGH DNLYWY RRVMGKEIKFLLHFVKESKQ DESGMPNNRFLAERTGGTYSTLKVQPAELEDSGVYFCASS
HBVT02	17.1	MSNQVLCCVVLCFLGANTV	DGGITQSPKYLFRKEGQNVTLSCEQNLNH DAMYWY RQDPGQGLRLIYYSQIVND FQKGDIAEGYSVSREKKESFPLTVTSAQKNPTAFYLCASS
HBVT56	18.1	MANSAMDTRVLCCAVICLLGAGLS	NAGVMQNPRHLVRRRGQEARLRCSPMKGH SHVYWY RQLPEEGLKFMVYLQKEN IIDESGMPKERFSAEFPKEGPSILRIQQVVRGDSAAYFCASS
PH29	18.2	MDTRLLCCAVICLLGAGLS	NAGVMQNPRHLVRRRGQEARLRCSPMKGH SHVYWY RQLPEEGLKFMVYLQKEN IIDESGMPKERFSAEFPKEGPSILRIQQVVRGDSAAYFCASS
HBVT72	19.1	CFVLFVKLFPGLFLQHLFLSLTDSF	HAKVTQTPGHLVKGKGQKTKMDCTPEKGH TFVYWY QQNQNKEFMLLISFQNEQ VLQETEMHKKRFSSQCPKNAPCSLAILSSEPGDTALYLCASS
HUT (HUT102)	20.1		TIHQWPATLVQPVGSPLSLECTVEGTSNPNLYWY RQAAGRGLQLLFYSIGIDQIS SEVPQNLSASRPQDRQFILSSKKLLLSDSGFYLCAWS
		← leader →	← V_B →

Figure 14. Deduced amino acid sequences of human TcR V_β segments. Spaces at equivalent positions in the nucleotide and deduced amino acid sequences of a given cDNA clone were added to maximize regions of homology. Introns have been removed from rearranged genomic sequences obtained from ATL cell lines (ATL21, ATL121, MT11) and CEM. Sequences shown in small capitalized letters were not used in the consensus calculation.

Table III. Human TcR Vβ and Jβ use in cDNA clones

a)

cDNA library	Number of unique cDNAs	Number of Vβ	Number of unique Vβ	Number using each Vβ family																				
				1	2	3	4	5	6	7	8	9	10	11	12	13[a]	14[a]	15	16	17	18	19	20
HBVP	13	10	9		1	1	1	1	3		1				2				1				
PL	27	26	18	4	2	4	2	3	2	2	1	1	1	1	3								
PH	15	15	12	2	2	1	1	1	4		2			1	1			1			2		
HBVT	11	11	8	2																1	1		
other	15	15	15	2	1	2	1	3	3		1	1	1		1					1		1	1

b)

cDNA library	Number of unique cDNAs	Number using Cβ1	Number using Cβ2	Number using each Jβ1 gene segment						Number using each Jβ2 gene segment						
				1	2	3	4	5	6	1	2	3	4	5	6	7
HBVP	13	4	8	2		1				2		3	1	1		1
PL	27	10	17	5			1	3	1	7	2	5		2	1	
PH	15	NR	NR	2		2	1			1		1	2	2	2	4
HBVT	11	3	8	1	2					4	1		1			2
other	15	5	5	2	2					2	2		1	1	3	2

[a] As a result of sequence comparison, the Vβ13 and Vβ14 families have been amalgamated into Vβ12 and Vβ3 families, respectively.

6. T-Cell Receptor Diversity

In summary, there is an extensive potential repertoire for the assembly of both α and β variable domains. The diversity of the receptor is the product of the α- and β-chain diversities. Purely from combinatorial joining, a diversity of over 10^6 can be created. In mice, $20(V_\beta) \times 2(D_\beta) \times 12(J_\beta) \times 50(V_\alpha) \times 50(J_\alpha) = 1.2 \times 10^6$ and in humans, $100(V_\beta) \times 2(D_\beta) \times 13(J_\beta \times 100(V_\alpha) \times 50(J_\alpha) = 1.3 \times 10^7$. If one assumes that the use of different reading frames, combined with the somatic mutation mechanisms of junctional diversity and N-region diversity, conservatively produce a hundredfold increase in the diversity of each chain, this increases the total receptor diversity to over 10^{10} different receptors (see Tables V and VI). Even allowing for mismatches between certain α and β chains, this should be enough to cope with the onslaught of foreign antigens.

There are a number of factors affecting the diversity of the T-cell receptor repertoire. For example, the second D_β, $D_{\beta 2}$, cannot recombine with the members of the first J_β cluster by the looping excision mechanism, but can by some of the alternate mechanisms. This decrease is balanced by the increase resulting from the possibilities of directly joining of V_β to J_β segments, of the use of more than one D_β to D_β joining, and of the reading of the D_β segments in any of the three reading frames. For example, the human $D_{\beta 2}$ encodes a stop codon in one reading frame, but it is still used in productive rearrangements, if the stop codon has been deleted in the rearrangement process (Concannon et al., 1986).

Somatic mutation also serves to increase β-chain diversity by means of the two basic mechanisms of junctional diversity and N-region diversity. Junctional diversity is the result of an imprecise joining between segments that can lead to the deletion of nucleotides from the rearranging ends of V_β, D_β, and J_β gene segments, leading to codon changes at the junctions (Siu et al., 1984b). N-region diversity is generated by the addition of nucleotides, that are not encoded by either of the rearranging segments, to their junctions during the rearrangement process. Both these processes substantially increase the diversity of T-cell receptor sequences, but also contribute to the generation of a large proportion of nonfunctional genes by the generation of shifts in reading frame.

Somatic hypermutation, which generates point mutations throughout the variable regions of immunoglobulin, does not seem to occur in either α- or β-chain sequences. This hypothesis is supported by the fact that comparison of the same V_β, which has associated with different D_β and J_β segments, in different T-cell clones, reveals no appreciable difference (Chien et al., 1984b; Patten et al., 1984; Barth et al., 1985; Goverman et al., 1985; Rupp et al., 1985; Kimura et al., 1986). The apparent lack of somatic hypermutation in both α and β chains could have any one of several explanations. First, it is possible that somatic hypermutation does occur, but it has not been detected since the majority of T-cell receptor sequences analysed have been derived either from relatively immature thymocytes or from T-cell lines that were only stimulated once in vivo and then passaged in vitro under conditions of target excess, which would not provide the selection pressure that has been shown to be necessary for

```
                                                              PERCENT
                                                              CONSENSUS
                                              F   GI            100
                                             EG  GI L           90
                                             EG  GI L V         80
                                             EG  GI L V P       70
                                           L EG  GI L V P       60
                      D A Y C              KL EQ  GI L V P       50
                      D A Y CA            KL EQ  GI L V P       40
                      D A Y CA            KL EQKGI L V P        30
                      DSA YFCA           KLEIFGKGTRLTV P        20
              JALPHA  DSAVYFCA          GGSGGKLIFGKGTRLTVIP     10
cDNA/GENOMIC CLONE    FAMILY  DSAVYFCA  VGGGSGGKLIFGKGTRLTVIP
```

cDNA/GENOMIC CLONE	JALPHA FAMILY		
GERMLINE	A		/GYSSASKIIFGSGTRLSIRP/
HAVP35		DSATYLCA	LAPSYSSASKIIFGSGTRLSIRP	NIQN
PGA (HPBMLT)		DSAVYFCA	LDSSASKIIFGSGTRLSIRP	NIQN
SUPT1B		DTAVVYCA	RVRRRYSSASKIIFGSGTRLSIR
HAVT18, PY14 (JURKAT)		DAAEYFCAVSDLEPNSSASKIIFGSGTRLSIRP		NIQN
GERMLINE	B		/MDSSYKLIFGSGTRLLVRP/
HAVP42			/MDSSYKLIFGSGTRLLVRP	HIQN
SUPT1A (SUPT1)		DSASYFCAVPPtg MDSSYKLIFGSGTRLLVA		
GERMLINE	C		/SSGSARQLTFGSGTQLTVLP/
GERMLINE	D		/TTDSWGKFEFGAGTQVVVTP/
GERMLINE	E		/EGQGFSFIFGKGTRLLVKP/
GERMLINE	F		/NSGNTPLVFGKGTRLSVIA/
DT55 (DT)		DTAVYYC	IVRAINSGNTPLVFGKGTRLSVIA	NIQN
HAVP49	G	DSAVYFCA	AKRKASSNTGKLIFGQGTTLQVKP	DIQN
HAVP29, HAVP32		DAAVYYC	GLPSNTGKLIFGQGTTLQVKP	DIQN
HAVP36	H	DSAMYYCA	LSVYNQGGKLIFGQGTELSVKP	NIQN
HAVT20		DAAMYYCA	IIDNQGGKLIFGQGTELSVKP	NIQN
HAJT23			NQGGKLIFGQGTELSVKP	NIQN
HAVP02	I	DAAVYYCA	VEVPNTDKLIFGTGTRLQVFP	NIQN
HAVP08	J	DAAVYYCI	RANAGGTSYGKLTFGQGTILTVHP	NIQN
HAVT06		DSATYLCA	VKPAGGTSYGKLTFGQGTILTVHP	NIQN
HAVP44	K	DTASYFCATPPLSSGGSNYKLTFGKGTLLTVNP		NIQN
HAVT33		DTAVYYCI	VRGNSGGSNYKLTFGKGTLLTVNP	NIQN
HAVP26	L	DSATYLCA	LRDGQKLLFARGTMLKVDL	NIQN
HAVP50	M	DSAVYFCA	EIGGEKLVFGQGTRLTINP	NIQN
HAVP10, HAVP60	N	DTAEYFCA	VNEYDYKLSFGAGTTVTVRA	NIQN
HAVP41, HAVP17	O	DSAVYFCA	ASRKDSGGYQKVTFGTGTKLQVIP	NIQN
HAVP25			GGYQKVTFGTGTKLQVIP	NIQN
HAVP05	P	DTASYFCA	TDGNRDDKIIFGKGTRLHILP	NI
HAVP71	Q	DSATYLCA	VNYPRGTTLGRLYFGRGTQLTVWP	DIQN
HAVT32, HAVT35		DSALYFCA	VRPDRGSTLGRLYFGRGTQLTVWP	DIQN
HAJP28	R		LRARNNARLMFGDGTQLVVKP	NI
HAVP21	S	DSASYFCA	VFNQAGTALIFGKGTTLSVSS	NIQN
HAVP58	T	DTGHYLCA	GVSSGGSYIPTFGRGTSLIVHP	YIQN
HAVT31		DSAIYFCA	ESKTPSRPTFGRGTSLIVHP	YIQN
HAVP51			IPTFGRGTSLIVHP	YIQN
HAVP01	U	DSAMYFCA	SREGSGNQFYFGTGTSLTVIP	NI
HAVT24		DSAVYFCA	EGPPTGNQFYFGTGTSLTVIP	NIQN
HAVT01	V	DTAVYYCI	ALYSGAGSYQLTFGKGTKLSVIP	NIQN
HAVT27	W	DTAVVYCI	VRDWVGGGADGLTFGKGTHLIIQP	YIQN
HAVT15		DSGVFCA	ALDLWGGADGLTFGKGTHLIIQP	YIQN
HAJT17	X	ctatgtg aagatcacctag MLNFGKGTELIVSL		DIQN
(GERMLINE VALPHA)		DTAEYFCAVSga cacagtg cctgagactgcaggagagctga acacaaacg		

```
                  ←Vα──→ ←Dα?→← ─────Jα─────→  | ←Cα→
```

somatic hypermutation in B cells. An alternative hypothesis is that while somatic hypermutation can occur in T cells, it is a rare event because of a lack of
selection pressure *in vivo* for the expansion of hypermutated cells as a result of a
lower dependence on the affinity for the antigen/MHC complex for T-cell activation than is the case for the affinity for antigen by immunoglobulin in B cells.
The third hypothesis is that somatic hypermutation is forbidden in T cells because it would allow the generation of potentially harmful specificities. T cells
undergo a rigorous selection procedure in the thymus that results in T-cell recognition of antigen in the context of self MHC, and the development of new,
unselected specificities after thymic export could generate a number of potentially dangerous autoreactive T-cell clones. Our present knowledge of T-cell
function is not extensive enough to determine which of these hypotheses is
valid, but it is hoped that the use of molecular biology to study the problems
raised by cellular immunology will provide answers to questions such as these in
the future.

7. Conclusion

The elusive and mysterious T-cell antigen-receptor genes have now been
cloned and largely characterized. It has been established that they are distinct
from the immunoglobulin genes, although their basic genomic organization and
mechanisms of recombination are similar to those of the B-cell antigen-recogni-
tion genes. The availability of these genes should enable some of the fundamental problems facing immunologists of the past few decades to be tackled. The
molecular basis of the dual recognition of antigen and MHC genes by T cells is
one such problem, which will most likely be attacked by the use of DNA transfer of TcR genes to reconstitute given T-cell recognition activities. Already the
experiments of Dembic *et al.* (1986) have reconstituted a T-cell antigen receptor
with a given specificity by transfer of only the α- and β-chains from a cell with
that specificity, providing strong evidence to support the one receptor model.
Another such problem is the elucidation of the mechanisms by which the thymus

Figure 15. Deduced amino acid sequences of human TcR J_α gene segments. Spaces at equivalent
positions in both the nucleotide and deduced amino acid sequences have been added to maximize
regions of homology. Sequences from cDNA/genomic clones used in Fig. 13 as well as six non-V_α-
containing cDNAs (HAVP25, HAVP42, HAVP51, HAJP28, JAJT17, HAJT23) and the six germline J_α segments are included (/, indicates the germline boundaries). Sequences shown in small
capitalized letters were not used in the consensus calculation. Small lower case letters denote nucleotide residues. In $J_\alpha X$ an in-frame stop codon is underlined and the heptamer of the putative recombination recognition signal is boxed. A germline V_α sequence is shown at the bottom for the
definition of V–J boundaries.

Table IV. Human TcR Vα and Jα Use in cDNA Clones

a)

cDNA library	Number of unique cDNA	Number of Vα	Number of unique Vα	8Vα family															
				1	2	3	4	5	6	7	8	9	10	11	12	13	14	15	16
HAVP	21	18	13	1	2	2	1	1	1	2	3	1	1	3					
HAVT	12	10	8	.1	1		3				1					1	1	1	1
Other	6	4	4	1		1			1					1				1	

b)

cDNA library	Number of Jα	Number of unique Jα	Jα gene segment																							
			A	B	C	D	E	F	G	H	I	J	K	L	M	N	O	P	Q	R	S	T	U	V	W	X
HAVP	20	17	1	1						2	1	1	1	1	1	1	1	2	1	1	1	2	1	1		
HAVT	12	10	1							2	1		1		1		1	1		1		1	1	1	2	1
Other	5	3	3		1			1																		

Table V. Estimate of Murine TcR Repertoire

Gene segment	α Chain	β Chain
V	50	20
D	?	$\leqslant 2$
J	50	12
C	1	2
Combinations	2.5×10^3	4×10^2
N sequences added	2.5×10^5	4×10^4
αβ combinations	approximately 10^{10}	

Table VI. Estimate of Human TcR Repertoire

Gene segment	α Chain	β Chain
V	50	100
D	?	$\leqslant 2$
J	50	13
C	1	2
Combinations	2.5×10^3	3.6×10^3
N sequences added	2.5×10^5	3.6×10^5
αβ combinations	approximately 10^{11}	

selects against self-reactive T cells and how T cells specific for self- restricted antigens are generated. An understanding of these processes involved in thymic education may be gained by the construction of transgenic mice with given T-cell receptor genes.

The cloning of the genes encoding these T-cell recognition structures have also opened up a number of avenues that may lead to a better understanding of certain clinical problems, such as the determination of the clonality and lineage of a given malignant hematopoetic and autoimmune disorders. Already, the lineage and clonality of a number of leukemias and lymphomas have been better defined as a result of the use of these genes.

References

Acuto, O., Hussey, R., Fitzgerald, K., Protentis, J., Meuer, S., Schlossman, S., and Reinherz, E., 1983, The human T cell receptor: Appearance in ontogeny and biochemical relationship of α and β subunits on IL-2 dependent clones and T cell tumors, Cell 34:717–726.

Acuto, O., Fabbi, M., Smart, J., Poole, C., Protentis, J., Royer H., Schlossman, S., and Reinherz, E.L., 1984, Purification and N-terminal sequencing of the β subunit of a human T cell antigen receptor, Proc. Natl. Acad. Sci. USA 81:3851–3855.

Alt, F.W., Yancopoulos, G.D., Blackwell, T.K., Wood, C., Thomas,E., Bos, M., Coffman, R., Rosenberg, N., Tonegawa, S., and Baltimore D., 1984, Ordered rearrangement of immunoglobulin heavy chain variable region segments, *EMBO J.* **3**:1209–1219.

Arden, B., Klotz, J.L., Siu, G., Hood, L., 1985, Diversity and structure of genes of the α family of mouse T cell antigen receptor, *Nature* **317**:783–787.

Barth, R., Kim, B., Lan, N., Hunkapiller, T., Sobieck, N., Winoto, A., Gershenfeld, H., Okada, C., Hansburg, D., Weissman, I., and Hood, L., 1985, The murine T-cell receptor employs a limited repertoire of expressed V_β gene segments, *Nature* **316**:517–523.

Becker, D.M., Patten, P., Chien, Y.-H., Yokota, T., Eshhar, Z., Giedlin, M., Gascoigne, N.R., Goodnow, C., Wolf, R., Arai, K., and Davis, M., 1985, Variability and repertoire size of T-cell receptor V_α gene segments, *Nature* **317**:430–434.

Behlke, M.A., Spinella, D.G., Chou, H.S., Sha, W., Hartl, D.L., and Loh, D.Y., 1985, T-cell receptor β-chain expression: dependence on relatively few variable region genes, *Science* **229**:566–570.

Bernstein, R., Pinto, M., and Jenkins, T., 1981, Ataxia telangiectasia with evolution of monosomy 14 and emergence of Hodgkin's Disease, *Cancer Gen. Cytogen.* **4**:31–37.

Biddison, W.E., Rao, P.E., Talle, M.A., Goldstein, G., and Shaw, S., 1984, Possible involvement of the T4 molecule in T-cell recognition of class II HLA antigens: evidence from studies of CTL-target cell binding, *J. Exp. Med.* **159**:793–797.

Born, W., Yague, J., Palmet, E., Kappler, J., and Marrack, P., 1985, Rearrangement of T-cell receptor β-chain genes during T cell development, *Proc. Natl. Acad. Sci. USA* **82**:2925–2929.

Borst, J., Alexander, S., Elder, J., Terhorst, C., 1983, The T3 complex on human T lymphocytes involves four structurally distinct glycoproteins, *J. Biol. Chem.* **258**:5135–5143.

Caccia, N., Bruns, G.A.P., Kirsch, I.R., Hollis, G.R., Bertness, V., and Mak, T.W., 1985, T cell receptor α chain genes are located on chromosome 14 at 14q11–14q12 in humans, *J. Exp. Med.* **161**:1255–1260.

Caccia, N., Kronenberg, M., Saxe, D., Haars, R., Bruns, G., Goverman, J., Malissen, M., Willard, H., Simon, M., Hood, L., and Mak, T.W., 1984a, The T-cell receptor β chain genes are located on chromosome 6 in mice and chromosome 7 in humans, *Cell* **37**:1091–1099.

Caccia, N., Mak, T.W., and Klein, G., c-*myc* Involvement in chromosomal translocations in mice and men, in: *Cellular and Molecular Biology of Neoplasia,* (T.W. Mak and I. Tannock, eds.), pp.199–209, Alan R. Liss, New York, 1984b.

Chang, H.-C., Seki, T., Moriuchi, T., and Silver, J., 1985, Isolation and characterization of mouse Thy-1 genomic clones, *Proc. Natl. Acad. Sci. USA* **82**:3819–3823.

Chien, Y., Becker, D., Lindsten, T., Okamura, M., Cohen, D., and Davis, M., 1984a, A third type of murine T cell receptor gene, *Nature* **312**:31–35.

Chien, Y.-H., Gascoigne, N.R.J., Kaveler, J., Lee, N.E., and Davis, M.M., 1984b, Somatic recombination in a murine T-cell receptor gene, *Nature* **309**:322–326.

Clark, M.J., Gagnon, J., Williams, A.F., and Barclay, 1985, MRC OX-2 antigen: a lymphoid/neuronal membrane glycoprotein with a structure like a single immunoglobulin chain, *EMBO J.* **4**:113–118.

Clark, S.P., Yoshikai, Y., Siu, G., Tayler, S., Hood, L., and Mak, T.W., 1984, Identification of a diversity segment of the human T-cell receptor beta chain, and comparison to the analogous murine element, *Nature* **311**:387–389.

Concannon, P., Pickering, L.A., Kung, P., Hood, L., 1986, Diversity and structure of human T-cell receptor β-chain variable region genes, *Proc. Natl. Acad. Sci. USA* **83**:6598–6602.

Cory, S., Adams, J.M., and Kemp, D.J., 1980, Somatic arrangements for forming active immunoglobulin γ genes in B and T lymphoid cell lines, *Proc. Natl. Acad. Sci. USA* **77**:4943–4947.

Davies, D.R., and Metzger, H., 1983, Structural basis of antibody function, *Annu. Rev. Immunol.* **1**:63–86.

Dembic, Z., Bannworth, W., Taylor, B.A., Steinmetz, M., 1985, The gene encoding the T cell receptor α-chain maps close to the Np-1 locus on mouse chromosome 14, *Nature* **314**:271–273.

Dembic, Z., Haas, W., Weiss, S., McCubrey, J., Kiefer, H., von Boehmer, H., and Steinmetz, M., 1986, Transfer of specificity by murine α and β T cell receptor genes, *Nature* **320:**323–328.

Duby, A.D., Klein, K.A., Murre, C., and Seidman, J.G., 1985, A novel mechanism of somatic rearrangement predicted by a human T cell antigen receptor β-chain complementary DNA, *Science* **228:**1204–1206.

Early, P., Huang, H., Davis, M., Calame, K., and Hood, L., 1981, An immunoglobulin heavy chain variable region gene is generated from three segments of DNA: V_H, D and J_H, *Cell* **19:**981–992.

Engleman, E.G., Benike, C., Glickman, E., and Evans, R.L., 1981, Antibodies to membrane structures that distinguish suppressor/cytotoxic and helper T lymphocyte subpopulations block the mixed leukocyte reaction in man, *J. Exp. Med.* **154:**193–198.

Fedderson, R.M., and Van Ness, B.G., 1985, Double recombination of a single immunoglobulin κ allele: Implications for the mechanism of rearrangement, *Proc. Natl. Acad. Sci. USA* **82:**4793–4797.

Forster, A., Hobart, M., Hengartner, H., Rabbitts, T.H., 1980, An immunoglobulin heavy-chain gene is altered in two T-cell clones, *Nature* **286:**897–899.

Gascoigne, N., Chien, Y., Becker, D., Kavaler, J., Davis, M., 1984, Genomic organization and sequence of T cell receptor β-chain constant and joining region genes, *Nature* **310:**387–391.

Goverman, J., Minard, K., Shastri, N., Hunkapiller, T., Hansburg, D., Sercarz, E., Hood, L., 1985, Rearranged β T-cell receptor genes in a helper T cell clone specific for lysozyme: no correlation between V_β and MHC restriction, *Cell* **40:**859–867.

Haars, R., Kronenberg, M., Owen, F., Gallatin, M., Weissman, I., and Hood, L., 1986, Rearrangement and expression of T-cell antigen receptor and ψ chain genes during thymic differentiation, *J. Exp. Med.* **164:**1–24.

Haskins, K., Kappler, J., and Marrack, P., 1984, The major histocompatability complex-restricted antigen receptor on T cells, *Annu. Rev. Immunol.* **2:**51–66.

Hayday, A.C., Diamond, D.J., Tanigawa, G., Heilig, J.S., Folsom, V., Saito, H., and Tonegawa, S., 1985, Unusual organization and diversity of T cell receptor α chain genes, *Nature* **316:**828–832.

Hecht, F., Morgan R., Hecht, B., and Smith, S.D., 1984, Common region on chromosome 14 in T-cell leukemia and lymphoma, *Science* **226:**1445–1447.

Hedrick, S.M., Germain, R.N., Bevan, M.J., Dorf, M., Engel, I., Fink, P., Gascoigne, N., Heber-Katz, E., Kapp, J., Kaufman, Y., Kaye, J., Melchers, F., Pierce, C., Schwartz, R.H., Sorensen, C., Taniguchi, M., and Davis, M.M., 1985, Rearrangement and transcription of a T-cell receptor β chain gene in different T-cell subsets, *Proc. Natl. Acad. Sci. USA* **82:**531–535.

Hedrick, S., Nielsen, E., Kavaler, J., Cohen, D., Davis, M., 1984, Sequence relationships between putative T-cell receptor polypeptides and immunoglobulins, *Nature* **308:**153–158.

Hochtl, J., Muller, C.R., and Zachau, H.G., 1982, Recombined flanks of the variable and joining segments of immunoglobulin genes, *Proc. Natl. Acad. Sci. USA* **79:**1383–1387.

Honjo, T., 1983, Immunoglobulin genes, *Annu. Rev. Immunol.* **1:**499–528.

Hozumi, N., and Tonegawa, S., 1976, Evidence for somatic rearrangement of immunoglobulin genes coding for variable and constant regions, *Proc. Natl. Acad. Sci. USA* **73:**3628–3632.

Ikuta, K., Ogura, T., Shimizu, A., Honjo, T., 1985, Low frequency of somatic mutation in β-chain variable region genes of human T-cell receptors, *Proc. Natl. Acad. Sci. USA* **82:**7701–7705.

Kappler, J., Kubo, R., Haskins, K., White, J., Marrack, P., 1983, The mouse T-cell receptor: Comparison of MHC-restricted receptors on two T-cell hybridomas, *Cell* **34:**727–737.

Kavaler, J., Davis, M.M., and Chien, Y.-H., 1984, Localization of a T cell receptor diversity-region element, *Nature* **310:**421–423.

Kemp, D.J., Adams, J.M., Mottram, P.L., Thomas W.R., Walker, I.D., and Miller, J.F.A.P., 1982, A search for messenger RNA molecules bearing immunoglobulin$_H$ nucleotide sequences in T cells, *J. Exp. Med.* **156:**1848–1853.

Kimura, N., Toyonaga, B., Yoshikai, Y., Triebel, F., Debre, P., Minden, M., and Mak, T.W.,

1986, Sequences and diversity of human T cell receptor β chain variable region genes, *J. Exp. Med.* **164:**739–750.

Kimura, N., Toyonaga, B., Yoshikai, Y., Du, R.P., and Mak, T.W., 1987, Sequences and repertoire of the human α and β chain variable region genes in thymocytes, *Eur. J. Immunol.*, (in press).

Kraig, E., Kronenberg, M., Kapp, J., Pierce, C.W., Abruzzini, A.F., Sorensen, C.M., Samelson, L.E., Schwartz, R.H., and Hood, L.E., 1983, T and B cells that recognize the same antigen do not transcribe similar heavy chain variable regions gene segments, *J. Exp. Med.* **158:**192–209.

Kronenberg, M., Davis, M.M., Early, P.W., Hood, L.E., and Watson, J.D., 1980, Helper and killer T cells do not express B cell immunoglobulin joining and constant region gene segments, *J. Exp. Med.* **152:**1745–1761.

Kronenberg, M., Kraig, E., Horvath, S.J., and Hood, L.E., 1982, Cloned T cells as a tool for molecular geneticists: Approaches to cloning genes which encode T-cell antigen receptors, in: *Isolation, Characterization and Utilization of T Lymphocyte Clones,* (C.G. Fathman and F. Fitch, eds.), pp. 467–491, Academic Press, New York.

Kronenberg, M., Kraig, E., Siu, G., Kapp, J.A., Kappler, K., Marrack, P., Pierce, C.W., and Hood, L., 1983, Three T-cell hybridomas do not express detectable heavy-chain variable gene transcripts, *J. Exp. Med.* **158:**210–227.

Kronenberg, M., Goverman, J., Haars, R., Malissen, M., Kraig, E., Phillips, L., Delovitch, T., Suciu-Foca, N., and Hood, L., 1984, Rearrangement and transcription of the β chain genes of the T cell antigen receptor in different types of murine lymphocytes, *Nature* **313:**647–653.

Kurosawa, Y., von Boehmer, H., Haas, W., Sakano, H., Trauneker, A., and Tanegawa, S., 1981, Identification of D segments of immunoglobulin heavy-chain genes and their rearrangement in T lymphocytes, *Nature* **290:**566–570.

Landegren, U., Romstedt, U., Axberg, I., Ullberg, M., Jondal, M., and Wigzell, H., 1982, Selective inhibition of human T-cell cytotoxicity at levels of target recognition or initiation of lysis by monoclonal OKT3 and Leu-2a antibodies, *J. Exp. Med.* **155:**1579–1584.

Larhammar, D., Gustafsson, K., Claesson, L., Winan, K., Schenning, L., Sundelin, J., Widmar, E., Peterson, P., and Rask, L., 1982, Alpha chain of HLA-DR transplantation antigens is a member of the same protein superfamily as the immunoglobulins, *Cell* **30:**153–161.

Lewis, S., Rosenberg, N., Alt, F., Baltimore, D., 1982, Continuing kappa-gene rearrangement in a cell line transformed by Abelson Murine Leukemia Virus, *Cell* **30:**807–816.

Lewis, S., Gifford, A., and Baltimore, D., 1984, Joining of $V_κ$ to $J_κ$ gene segments in a retroviral vector introduced into lymphoid cells, *Nature* **308:**425–428.

Littman, D.R., Thomas, Y., Maddon, P.J., Chess, L., Axel, R., 1985, The isolation and sequence of the gene encoding T8: A molecule defining functional classes of T lymphocytes, *Cell* **40:**237–246.

Maddon, P.J., Littman, D.R., Godfrey, M., Maddon, D.E., Chess, L., Axel, R., 1985, The isolation and nucleotide sequence of a cDNA encoding the T cell surface protein T4: A new member of the immunoglobulin gene family, *Cell* **42:**93–104.

Malissen, M., Hunkapiller, T., and Hood, L., 1983, Nucelotide sequence of a light chain of mouse I-A subregion: Aβ^d, *Science* **221:**750–754.

Malissen, M., McCoy, C., Blanc, D., Trucy, J., Devaux, C., Schmitt-Verhulst, A., Fitch, F., Hood, L., and Malissen, B., 1986, Direct evidence for chromosomal inversion during T-cell receptor β-gene rearrangements, *Nature* **319:**28–33.

Malissen, M., Minard, K., Mjolsness, S., Kronenberg, M., Goverman, J., Hunkapiller, T., Prystowsky, M.B., Fitch, F., Yoshikai, Y., Mak, T.W., and Hood, L., 1984, Mouse T-cell antigen receptor: Structure and organization of constant and joining gene segments encoding the β polypeptide, *Cell* **37:**1101–1110.

Marrack, P., Endres, R., Shimonkevitz, R., Zlotnik, A., Dialynis, D., Fitch, F., and Kappler, J., 1983, The major histocompatibility complex-restricted antigen receptor on T-cells. II. Role of the L3T4 product, *J. Exp. Med.* **158:**1077–1091.

Mathieu-Mahul, D., Caubet, J.F., Bernheim, A., Mauchaffe, M., Palmer, E., Berger, R., Larsen, C.-J., 1986, Molecular cloning of a DNA fragment from human chromosome 14(14q11) involved in T-cell malignancies, *EMBO J.* **4**:3427–3433.

McIntyre, B., and Allison, J., 1983, The mouse T cell receptor: Structural heterogeneity of molecules of normal T cells defined by Xenoantiserum, *Cell* **34**:739–746.

Meuer, S.C., Cooper, D.A., Hodgdon, J.C., Hussey, R.E., Fitzgerald, K.A., Schlossman, S., and Reinherz, E.L., 1983, Identification of the receptor for antigen and major histocompatiblity complex on human inducer T lymphocytes, *Science* **222**:1239–1242.

Mostov, K.E., Friedlander, M., and Blobel, G., 1984, The receptor for transepithelial transport of IgA and IgM contains multiple immunoglobulinlike domains, *Nature* **308**:37–43.

Nakanishi, K., Sugimura, K., Yoaita, Y., Maeda, K., Kashiwamura, S.-I., Honjo, J., Kishimoto, T., 1982, A T15-idiotype positive T suppressor hybridoma does not use the T15 V_H gene segment, *Proc. Natl. Acad. Sci. USA* **79**:6984–6988.

Nakauchi, H., Nolan, G.P., Hsu, C., Huang, H.S., Kavathas, P., Herzenberg, L.A., 1985, Molecular cloning of Lyt-2, a membrane glycoprotein marking a subset of mouse T lymphocytes: Molecular homology to its human counterpart, Leu-2/T8, and to immunoglobulin variable regions, *Proc. Natl. Acad. Sci. USA* **82**:5126–5130.

Ohashi, P., Mak, T.W., Van den Elsen, P., Yanagi, Y., Yoshikai, Y., Calman, A.F., Terhorst, C., Stobo, J.D., and Weiss, A., 1985, Reconstitution of an active T3/T cell antigen receptor in human T cells by DNA transfer, *Nature* **316**:602–606.

Patten, P., Yokota, T., Rothbard, J., Chien, Y.-H., Arai, K.-I., and Davis, M.M., 1984, Structure, expression and divergence of T-cell receptor β-chain variable regions, *Nature* **312**:40–46.

Raulet, D.H., Garman, R.D., Saito, H., Tonegawa, S., 1985, Developmental regulation of T-cell receptor gene expression, *Nature* **314**:103–107.

Rupp, F., Acha-Orbea, H., Hengartner, H., Zinkernagel, R., Joho, R., 1985, Identical $v_β$ T-cell receptor genes used in alloreactive cytotoxic and antigen plus I-A specific helper T-cells, *Nature* **315**:425–427.

Saito, H., Kranz, D.M., Takagaki, Y., Hayday, A., Eisen, H., Tonegawa, S., 1984, A third rearranged and expressed gene in a clone of cytotoxic T lymphocytes, *Nature* **312**:36–40.

Saito, T., Weiss, A., Miller, J., Norcross, M.A., Germain, R.G., 1987, Specific antigen-Ia activation of transfected human T cells expressing murine Ti αβ-human T3 receptor complexes, *Nature* **325**:125–130.

Sangster, B., Minowada, J., Suci-Foca, N., Minden, M., and Mak, T.W., 1986, Rearrangement and expression of the α, β and γ T cell receptor genes in human leukemias and functional T cells, *J. Exp. Med.* **163**:1491–1507.

Selsing, E., and Storb, U., 1981, Mapping of immunoglobulin variable region genes: relationship to the "deletion" model of immunoglobulin gene rearrangement, *Nucl. Acids. Res.* **9**:5725–5735.

Seki, T., Spurr, N., Obata, F., Goyert, S., Goodfellow, P., Silver, Jr., 1985, The human Thy-1 gene: Structure and chromosomal location, *Proc. Natl. Acad. Sci. USA* **82**:6657–6661.

Shima, E.A., Le Beau, M.M., McKeithan, T.W., Minowada, J., Showe, L.W., Mak, T., Rowley, J.D., and Diaz, M.O., 1986, T-cell receptor α-chain moves immediately downstream of c-myc in a chromosomal 8;14 translocation in a cell line from a human T cell leukemia, *Proc. Natl. Acad. Sci. USA* **83**:3439–3443.

Sim, G., Yague, J., Nelson, J., Marrack, P., Palmer, E., Augustin, A., and Kappler, J., 1984, Primary structure of human T cell receptor α chain, *Nature* **312**:771–775.

Siu, G., Clark, S., Yoshikai, Y., Malissen, M., Yanagi, Y., Strauss, E., Mak, T.W., Hood, L., 1984a, The human T cell antigen receptor is encoded by variable, diversity, and joining gene segments that rearrange to generate a complex V gene, *Cell* **37**:393–401.

Siu, G., Kronenberg, M., Straus, E., Haars, R., Mak, T.W., and Hood, L., 1984b, The structure, rearrangement and expression of $D_β$ gene segments of the murine T cell antigen receptor, *Nature* **311**:344–350.

Steinmetz, M., Altenberger, W., Zachau, H.G., 1980, A rearranged DNA sequence possibly related to the translocation of immunoglobulin gene segments, *Nucl. Acids Res.* **8:**1709–1720.

Steinmetz, M., Frelinger, J.G., Fisher, D., Hunkapiller, T., Periera, P., Weissman, S.M., Vehara, H., Nathenson, S., Hood, L., 1981, Three cDNA clones encoding mouse transplantation antigens: Homology to immunoglobulin genes, *Cell* **24:**125–134.

Strominger, J.L., Orr, H.T., Parham, P., Ploegh, H.L., Mann, D.J., Bilofsky, Y., Saroff, H.A., Wu, T.T., Kabat, E.A., 1980, An evaluation of the significance of amino acid sequence homology in human histocompatibility antigens (HLA-A and HLA-B) with immunoglobulins and other proteins using relatively short sequences, *Scand. J. Immunol.* **11:**573–593.

Sukhatme, V.P., Sizer, K.C., Vollmer, A.C., Hunkapiller, T., Parnes, Jr., 1985, The T cell differentiation antigen Leu-2/T8 is homologous to immunoglobulin and T cell receptor variable regions, *Cell* **40:**591–597.

Swain, S.L., 1981, Significance of Lyt phenotypes Lyt2 antibodies block activities of T-cells that recognize class 1 major histocompatiblity complex antigens regardless of their function, Proc. *Natl. Acad. Sci. USA* **78:**7101–7105.

Tonegawa, S., 1983, Somatic generation of antibody diversity, *Nature* **302:**575–581.

Toyonaga, B., Yanagi, Y., Suciu-Foca, N., Minden, M.D., Mak, T.W., 1984, Rearrangement of the T cell receptor gene YT35 in human DNA from thymic leukemic T cell lines and functional helper, killer and suppressor T cell clones, *Nature* **311:**385–387.

Toyonaga, B., Yoshikai, Y., Vadasz, V., Chin, B., and Mak, T.W., 1985, Organization and sequences of the diversity, joining and constant region genes of the human T cell receptor β chain, *Proc. Natl. Acad. Sci. USA* **82:**8624–8628.

van den Elsen, P., Shepley, B., Borst, J., Coligan, J.E., Markham, A.F., Orkin, S., and Terhorst, C., 1984, Isolation of cDNA clones encoding the 20K T3 glycoprotein of the human T cell receptor complex, *Nature* **312:**413–417.

Van Ness, B.G., Coleclough, C., Perry, R.P., and Weigert, M., 1982, DNA between variable and joining segments of immunoglobulin κ light chain is frequently retained in cells that retain the κ locus, *Proc. Natl. Acad. Sci. USA* **79:**262–266.

Weiss, A., Stobo, J.D., 1984, Requirement for the coexpression of T3 and the T cell antigen receptor on a malignant T cell line, *J. Exp. Med.* **160:**1284–1299.

Williams, A.F., and Gagnon, J., 1982, Neuronal cell glycoprotein: Hemology with immunoglobulin, *Science* **216:**696–703.

Williams, D.L., Look, A.T., Melvin, S.L., Roberso, P.K., Dahl, G., Flake, T., and Stass, S., 1984, New chromosomal translocations correlate with specific immunophenotypes of childhood acute lymphoblastic leukemia, *Cell* **36:**101–109.

Winoto, A., Mjolsness, S., and Hood, L., 1985, Genomic organization of genes encoding mouse T cell receptor α chain, *Nature* **316:**832–836.

Wu, T.T., and Kabat, E.A., 1970, An analysis of the sequences of the variable regions of Bence Jones proteins and myeloma light chains and their implications for antibody complimentarity, *J. Exp. Med.* **132:**211–250.

Yanagi, Y., Chan, A., Chin, B., Minden, M., and Mak, T.W., 1985, Analysis of cDNA clones specific for human T cells and the α and β chain of the T cell receptor heterodimer from a human T cell line, *Proc. Natl. Acad. Sci. USA* **82:**3430–3434.

Yanagi, Y., Yoshikai, Y., Leggett, K., Clark, S., Aleksander, I., and Mak, T.W., 1984, A human T cell-specific cDNA clones encodes a protein having extensive homology to immunoglobulin chains, *Nature* **308:**145–149.

Yancopooulos, G.D., Blackwell, T.K., Suh, H., Hood, L., and Alt, F., 1986, Introduced T cell receptor variable regions gene segments recombine in pre-B cells: Evidence that B and T cells use a common mechanism, *Cell* **44:**251–259.

Yoshikai, Y., Anatoniou, A., Clark, S.P., Yanagi, Y., Sangster, R., Elsen, P., Terhorst, C., and Mak, T.W., 1984, Sequence and expression of two distinct human T cell receptor β chain genes, *Nature* **312:**521–524.

Yoshikai, Y., Clark, S.P., Taylor, S., Sohn, V., Wilson, B., Minden, M., and Mak, T.W., 1985, Organization and sequences of the variable, joining, and constant region genes of the human T cell receptor α chain, *Nature* **316:**837–840.

Yoshikai, Y., Kimura, N., Toyonaga, B., and Mak, T.W., 1986, Sequences and repertoire of human T cell receptor α chain variable regions in mature T lymphocytes, *J. Exp. Med.* **164:**90–103.

Zech, L., Gahrton, L., Hammarstrom, L., Juliusson, G., Mellstedt, H., Robert, K.H., and Smith, C.I.E., 1984, Inversion of chromosome 14 marks human T-cell chronic lymphocytic leukemia, *Nature* **308:**858–860.

Zinkernagel, R.M., and Doherty, P.C., 1975, H-2 compatibility requirement for T cell mediated lysis of target infected with lymphocytic choriomengitis virus. Different cytotoxic T cell specificities are associated with structures in H-2K or H-2D, *J. Exp. Med.* **141:**1427–1437.

Zuniga, M.C., D'Eustachio, P., and Ruddle, N., 1982, Immunoglobulin heavy-chain gene rearrangement and transcription in murine T cell hybrids and T lymphomas, *Proc. Natl. Acad. Sci. USA* **79:**3015–3019.

3

Immunoglobulin-Related Structures Associated with Vertebrate Cell Surfaces

A. NEIL BARCLAY, PAULINE JOHNSON, GEOFF W. McCAUGHAN, and ALAN F. WILLIAMS

1. Members of the Ig Superfamily

Table I lists structures that can be confidently regarded as being related to Ig in evolution along with the chromosome assignment of their genetic loci in man and mouse. The structures are diverse in their functions and cellular expression and are usually unlinked in the genome despite some notable exceptions. Models for most of the molecules in Table I are given in Fig. 1 in terms of segments that can be considered to be related to Ig domains. In this chapter we will restrict detailed discussion to molecules other than Igs, TcR, and MHC antigens and will consider only protein sequences in assessing relationships. The initial manuscript has been revised to include new sequences.

2. Igs and the Domain Hypothesis

   The possibility that all Ig chains were derived in evolution from a primordial sequence of about 100 amino acids was suggested on the basis of compari-

Abbreviations: β_2-m, β_2-microglobulin; Ig, immunoglobulin; MAb, monoclonal antibody; MHC, major histocompatibility complex; Tcr, T lymphocyte antigen receptor.

A. NEIL BARCLAY, PAULINE JOHNSON, GEOFF W. McCAUGHAN, and ALAN F. WILLIAMS ● MRC Cellular Immunology Research Unit, Sir William Dunn School of Pathology, University of Oxford, Oxford OX1 3RE, United Kingdom. *Present address for P.J.:* Cancer Biology Laboratory, The Salk Institute, La Jolla, California 92112. *Present address for G.W.McC.:* A.W. Morrow Gastroenterology and Liver Center, Royal Prince Albert Hospital, Camperdown, 2050, NSW Australia.

Table I. Molecules in the Ig Superfamily

Category	Amino acid sequence number and chain size on SDS PAGE ($M_r \times 10^{-3}$)	Chromosome assignment	
		Human	Mouse
Immunoglobulins (Igs)			
H chains (e.g., μs)	590:gp75	14	12
L chain kappa	214:gp25	2	6
L chain lambda	213:gp25	22	16
T cell receptor (TCR) and CD3 chains			
TcR α chain	250:gp43	14	14
β chain	282:gp43	7	6
γ chain	286:—	7	13
CD3 γ chain	160:gp26	11q23	—
δ chain	150:gp21	11q23	9
ε chain	185:p19	11q23	9
Major histocompatibility (MHC) antigens			
Class I H-chain	339:gp45	6	17
β_2 − m	99:p12	15	2
Class II α	229:gp28	6	17
β	229:gp32	6	17
β_2 − m associated antigens			
TL H chain	336:gp45	6	17
Qa H chain	313:gp47	6	17
CD1 H chain	—:gp43-49	1	—
T-marker antigens			
CD2	322:gp50	—	—
CD4	435:gp55	12	—
CD8 chain 1	210:gp32	2	6
chain 2 (rodents)	187:gp37	—	6
Lymphoid/brain associated antigens			
Thy-1	111:gp25	11q23	9
MRC OX-2	248:gp45	3	—
Immunoglobulin receptors			
Poly Ig R	755:gp90-95	—	—
Macrophage FcR	351:gp47-60	—	1
Neural-associated molecules			
1B236	626:gp100	—	—
N-CAM	—:gp120	11q23	9
	—:gp140		
	—:gp170		
Po myelin protein	219:gp30	—	—

Table I. (Continued)

Category	Amino acid sequence number and chain size on SDS PAGE ($M_r \times 10^{-3}$)	Chromosome assignment	
		Human	Mouse
Growth factor receptors			
PDGF receptor	1067:gp180	5q23q31	
CSF-1 receptor	953: —	5q33q34	

[a]The chain sizes on SDS PAGE are approximate values from references given in the text. The amino acid sequence numbers are for the processed polypeptide chain and are from (Kabat *et al.*, 1983) unless other refs are given. Species are as follows: Igs, mouse, Tcr, mouse α (Chien *et al.*, 1984), β (Saito *et al.*, 1984), γ (Saito *et al.*, 1984); MHC antigens, human; TL, mouse (Fisher *et al.*, 1985); Qa, mouse (Devlin *et al.*, 1985); CD4, human (Maddon *et al.*, 1985); CD8, rat (Johnson and Williams, 1986); Thy-1 and MRC OX-2, rat (Clark *et al.*, 1985); Poly Ig R, rabbit (Mostov *et al.*, 1984); CD3, human (Krissansen *et al.*, 1986; van den Elsen *et al.*, 1984; Gold *et al.* 1986); FcR, mouse (Ravetch *et al.*, 1986); CD2, rat (Williams *et al.*, 1987); 1B236, rat (Lai *et al.*, 1987); N-CAM, chicken, partial sequence published (Hemperley *et al.*, 1986). Po myelin protein, rat (Lemke and Axel, 1985); PDGF receptor, human (Yarden *et al.*, 1986); CSF-1 receptor, human (Coussens *et al.*, 1986). Human chromosome assignments are given in McAlpine *et al.*(1985) except for CD8 (Sukhatme *et al.*, 1985), MRC OX-2 (McCaughan *et al.*, 1987b), CD3 (Gold *et al.*, 1987b). Mouse chromosome assignments are in ref (Kabat *et al.*, 1983) except for Tcr (Kranz *et al.*, 1985), CD8 (Sukhatme *et al.*, 1985), Thy-1 (Giguere *et al.*, 1985), N-CAM (Nguyen *et al.*, 1986) and FcR (Ravetch *et al.*, 1986).

sons between the sequences of Ig L chains and of H-chain Fc fragments (Hill *et al.*, 1966). The scheme proposed by Hill *et al.* is shown in Fig. 2. Subsequent sequences of Ig chains supported these ideas (Edelman, 1970) and the determination of three-dimensional structures for Ig L chains and Fab and Fc fragments (Amzel and Poljak, 1979) proved the domain concept to be correct at the structural level. Of particular note was the finding that V and C domains had similar β-strand folding patterns since sequence similarities between these domains were below the threshold of significance in early statistical analyses (Edelman, 1970; Amzel and Poljak, 1979).

Although the V- and C-domain folding patterns are similar, they can be distinguished because V domains have an extra loop of sequence in the middle of the domain. This is seen in Fig. 3, which shows the pattern of β-strands in Ig domains and the position of the conserved disulphide bond, between β-strands B and F, which is usually a characteristic of the Ig-fold. V and C domains can also be distinguished because they show different patterns of conserved residues superimposed on the conserved sequences common to most domains (Williams *et al.*, 1985). One region showing patterns correlating with domain type is centered on the COOH-terminal Cys of the conserved disulphide bond and some alignments in this region are shown in Fig. 4. The pattern Asp X Gly or Ala X Tyr X Cys is almost invariant in Ig V-domains and common in V-related sequences whereas the pattern Tyr X Cys X Val X His or Asn is commonly seen in Ig C-domains and MHC Ig-related domains. In the V-domains the conserved Asp correlates with a conserved Arg or Lys residue at the base of β-strand D (Fig. 3) and in V-domain structures the Arg and Asp residues form a salt bridge (Williams *et al.*, 1985).

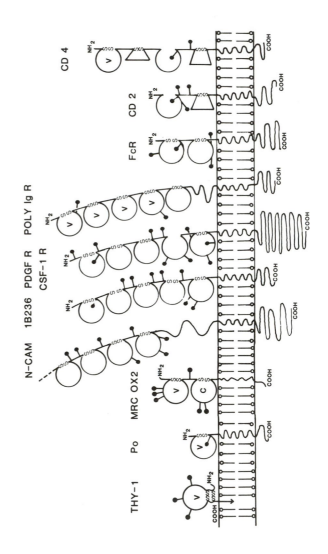

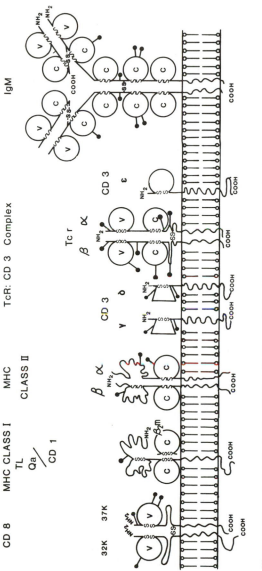

Figure 1. Shows models for the Ig-related molecules attached to a lipid bilayer. The circles show segments that are related to Ig-domains and the V or C inside the circle indicates a V-domain or C-domain or a segment that is clearly related to either of these domain types. Segments that are likely to be Ig-related but with considerable divergence from standard patterns are indicated by a rectangle. Interchain disulphide bonds are shown by (S-S). Postulated or established intrachain disulphide bonds are shown by (§) and the absence of those in domains of CD2, CD4, PDGF receptor and CSF-1 receptor indicates Ig-related segments without the conserved disulphide bond. Sites for N-linked glycosylation are indicated by (↑). Structures are for molecules from the following species and references are in the text: Thy-1, rat; MRC OX-2, rat; N-CAM, chicken; 1B236, rat; CD4, human; CD2, rat; FcR, mouse; Poly Ig R, rabbit; CD8, rat; MHC Class I and II, human; Tcr, mouse; CD3, human; IgM, human.

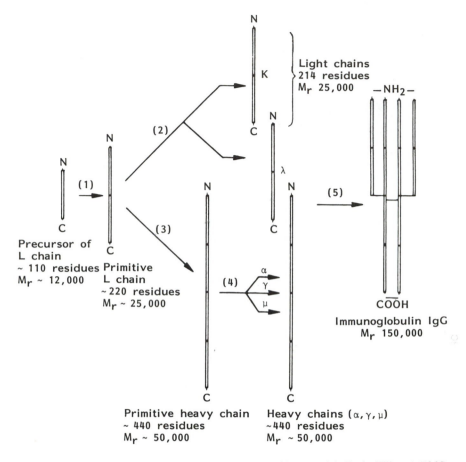

Figure 2. The original scheme proposed for the evolution of immunoglobulins by Hill *et al.* (1966).

Many of the Ig-related domains conform to a V or C domain in terms of both their likely folding pattern and the conserved pattern of sequences as shown in Fig. 4. This is the case for all domains labeled as V and C in Fig. 1. However, recently numerous exceptions to these correlations have been found. Ig-like domains in 1B236, N-CAM, α_1 B-glycoprotein, PDGF receptor, CSF-1 receptor, CD3 γ, δ and ϵ chains, and the macrophage Fc receptor are likely to have folding patterns like C domains yet they show conserved sequences typical of V domains around the COOH-terminal Cys residues. A more striking divergence from standard patterns is the possibility that domains can lack the conserved disulphide bond that has been considered an invariant feature of Ig-related se-

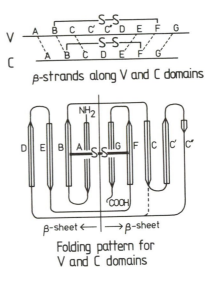

Figure 3. Folding patterns for Ig V and C domains. The upper part shows the disposition of β-strands marked by letters along V and C domains with the dashed lines indicating the β-strands which form the core of the fold that is common to the two domain types. The lower part shows a schematic view of the folding pattern for an Ig V-domain. The two β-sheets that form a sandwich held together by the conserved disulphide bond between β-strands B and F are laid out to show the fold. The C-domain fold is indicated by following the dashed line from β-strand C directly into β-strand D (adapted from Amzel and Poljak, 1979).

quences. One V_H domain from a functional antibody has been found to have a Tyr residue replacing the Cys that would normally form the COOH-terminal part of the conserved disulphide bond (Rudikoff and Pumphrey, 1986) and Ig-related domains lacking both Cys residues have been suggested to occur in the sequences of CD2 and CD4 T-cell antigens (Clark *et al.*, 1987; Williams *et al.*, 1987) (also domains 4 of PDGF receptor and CSF-1 receptor are Ig-related but lack Cys residues—see Section 4.8). In the case of the CD2 sequence there are a number of identities with conserved residues of V-like domains in the region of β-strands E and F but there is no Cys residue to form the centre of β-strand F. Instead Val 78 of CD2 is aligned with the conserved COOH-terminal Cys residue. This can be seen in Fig. 5, which shows alignments between domains I for human and rat CD2 antigen and domain I of human CD4 antigen which is V-like. The group of identities on the NH_2-terminal side of the second Cys is obvious in Fig. 5.

3. Statistical Significance of the Similarities to Ig

The statistical significance of an evolutionary relationship between proteins can at present only be tested via sequence identities and conservative substitutions. The Ig-related sequences are a divergent family and identities in the range 20–30% are the norm for most of the convincing comparisons between members of this group. When these similarities are assessed by eye, much significance is attached to conserved patterns of sequence (Williams *et al.*, 1985; Williams,

```
C-LAMBDA NEW        G I P D R I S A - - S K S G T S A T L A I T G L R T G D E A D Y Y C A T W D S S
V-KAPPA EU          G V P S R F I G - - S G S G T E F T L T I S S L Q P D D F A T Y Y C Q Q Y N S D
VH NEWM             P L R S R V T M L V D T S K N Q F S L R L S S V T A A D T A V Y Y C A R N L I A
VH EU               K F Q G R V T I T A D E S T N T A Y M E L S S L R S E D T A F Y F C A G G Y G I
T RECEPTOR          M P E D R F S A - K M P N A S F S T L K I Q P S E P R D S A V Y F C A S S F S T
RAT THY-1           T Y R S R V N L - - F S D R F I K V L T L A N F T T K D E G D Y M C E L R V S G
RAT MRC OX-2 (I)    T Y K D R I N I T E L G L L N T - S I T F W N T T L D D D E G C Y M C L F N M F G
RAB POLY IG R (I)   E Y S G R G K L T D F P D K G E F V V T V D Q L T Q N D S G S Y K C G V G V N G
RAB POLY IG R (IV)  D Y T G R L A L F E E P G N G T F S V V L N Q L T A E D E G F Y W C V S D D D E

                    ←──── BETA-STRAND D ────→
```

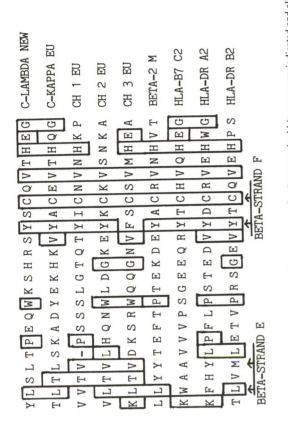

Figure 4. Alignments around β-strand F in V and C-type domains. All sequences are human except for the rat and rabbit sequences indicated and all are from Kabat *et al.*, (1983) except for poly Ig R sequences (Mostov *et al.*, 1984), T receptor sequences (Yanagi *et al.*, 1984) and MRC OX-2 domain I sequence (Clark *et al.*, 1985). Alignments for β-strands are as for V and C-domains from Amzel and Poljak (1979). Residues are boxed where ³⁄₉ have the same residue at any position in the V or C-domain set. Alignments are from Williams, *et al.* (1985).

```
         1         5          10          15    *    20          25
HUMCD4  -  N  K  V [V] L  G  K  K [G] D  T  V  E [L] T  C  T  A  S [Q] K  K  S  -
RATCD2  D  S  G  T [V] W  G  A  L [G] R  G  I  N [L] N  I  P  N  F [Q] M  T  D  D
HUMCD2  N  A  L  E  T  W [G] A  L [G] Q  D  I  N [L] D  I  P  S  F [Q] M  S  D  D

              30          35          40          45          50
HUMCD4 [I] Q  -  F  H [W] K  N  - [S] N  Q  I [K  I] L  G  N  Q  G  S [F] L  T [K] G  P  S
RATCD2 [I] D  E  V  R [W] E  R  G [S] -  -  -  T  L  V  A  E  -  -  - [F] K  R [K] M  K  P
HUMCD2 [I] D  D  I  K [W] E  K  T [S] D  K  K [K  I] -  A  Q  -  -  - [F] R  - [K] E  K  E

           55          60          65          70          75          80
HUMCD4  K [L] N  D  R  A [D] S  R  R  S [L] W  D  Q [G] N  F  P [L] I [I] K  N [L] K [I] E
RATCD2  F [L] -  -  K  S  G  A  F  E  - [I] L  A  N  G [D] -  - [L] K [I] K  N [L] T  R  D
HUMCD2  T  F  K  E  K  - [D] T  Y  K  - [L] F  K  N  G [T] -  - [L] K [I] K  H [L] K [T] D

                 85     *     90          95          100         105
HUMCD4 [D] S [D] T [Y] I  C  E [V] E [D] Q [K] E  E  V  Q  L [L] V  F  G [L] T  A  N  S  -
RATCD2 [D] S [G] T [Y] N  V  T [V] Y  S  T  N  G  T  -  R  I  L [D] K  A [L] D  L  R  I  L
HUMCD2 [D][Q][D] I [Y] K  V  S  I  Y [D] T [K] G  K  -  N  V [L] E  K  I  F  D  L  K  I  Q
```

Figure 5. Alignments for domains I of human and rat CD2 with human CD4 domain I. The alignments are similar to those produced by the ALIGN program for each of the CD2 sequences compared with CD4 with some adjustments to take into account matches between the three sequences. Residues aligned are: Human CD4 residues, 3-106; human CD2, 5-103; rat CD2, 2-98.

1984) (e.g., as shown in Fig. 4) and these are largely derived from amino acids that form the β-strands B, C, D, E, and F, which make up the core of the Ig-fold (Fig. 3).

In contrast to this, with the currently available statistical tests, similarities between sequences are scored regardless of whether or not they form part of a pattern that is characteristic of a superfamily. Thus sometimes similarities in sequence will be undervalued in that the pattern is there but by chance the coincidence of identities may be insufficient for a good score. In other cases two sequences might by chance give a score suggesting a relationship, yet this may be dubious because the conserved sequence patterns are absent. One way to get over this problem is to use a standard test but to consider scores for a set of Ig-related sequences that are divergent amongst themselves. If a test sequence is related to the set of sequences, then the scores taken as a whole should point to a relationship. In contrast a chance good score would not be supported by other comparisons in the set.

In Tables IIA–D sets of scores are shown for comparisons of some domains in the Ig superfamily tested by use of the ALIGN program (Dayhoff *et al.*, 1983). The scores are given in SD units (see legend to Table II) and some typical relationships between probability of a chance occurrence and an SD score are: probability 10^{-3}:SD, score 3.09; 10^{-5}:4.26; 10^{-7}:5.2; 10^{-9}:6.0; 10^{-15}:7.94.

Dayhoff *et al.* (1983) consider a score of 3 as a cut-off for indicating the possibility of an evolutionary relationship.

Table IIA shows comparisons amongst V and V-related domains together with some scores for unrelated sequences and scores for an Adenovirus sequence, which is discussed below. The control sequences are ovalbumin, ribonuclease, and superoxide dismutase and none of these show sequence similarities to Ig that are noticeable to the eye. However ribonuclease is similar to Ig in having a high content of β-structure and superoxide dismutase also has this feature, plus a folding pattern with some similarities to the Ig-fold (Richardson *et al.*, 1976). The set of values for the control sequences show that an occasional score approaching 3 SD can be obtained by chance but the whole set of data would provide no indication of evolutionary relationships. In contrast the pooled data for all the V and V-related sequences provides a strong case for arguing that these are a related family. It can be noted that the receptor V domains are more closely related to each other than to the other sequences with the exception of the V-like domain from CD8 chain II, which gives scores with V-domains at about the same level as seen within the V-domain set.

Table IIB shows comparisons amongst C and C-related domains and within this set the scores are also very good. The MRC OX-2 C-like segment gives the weakest set of scores in Table IIB, yet, taken together, the data for OX-2 seems convincing in comparison with the control scores.

The scores for comparisons between the V and C domains are considerably lower than those seen within each set (Table IIC) and the argument that these sequences are related was first clinched when the tertiary structures of Ig V and C domains were solved. One problem in comparing V and C-domain sequences is the considerable difference in length in the middle of the domain. However, even though the scores are weak some domains provide a good case for linking the V- and C-domain sets and one example is the MHC class II β-chain domain, which gave scores of 3 SD or more in five out of 13 comparisons with the V-domain set.

In contrast to this when the Adenovirus E3 glycoprotein sequence, which has been postulated to be Ig-related, was subjected to comparison with the set of Ig-related domains, no support was provided for a relationship (Chatterjee and Maizel, 1984). Over a sequence distance of 140 amino acids, the E3 sequence gave an alignment with one class II MHC sequence that produced a good score in a statistical test similar to the ALIGN program (Chatterjee and Maizel, 1984) and in a repeat using ALIGN we obtained a score of 4.2 SD units. Within this 140-residue segment, there were two Cys residues similarly placed to the conserved Cys residues in Ig-related domains and a segment of about 100 residues, including this Cys pair, was suggested as a possible Ig-related domain (Chatterjee and Maizel, 1984). However when this segment was included in the ALIGN comparisons in Table IIA and B, no scores of 3 S.D. units or more were seen and out of 26 comparisons with the V- or C-domain categories only 3 scores

Table II. Align Scores for Alignment of Ig V- and C-Domains and Related Sequences[a]

A

	Igλ	IgH	Igκ	TcBβ	TcRα	TcRγ	CD8R1	CD8R2	CD4	PIgRI	PIgRIII	OX-2	Thy-1	Adeno
		Receptor V domains[b]					V-related sequences[b]							
Igλ		7.1	12.9	11.0	7.3	9.0	5.1	9.5	4.5	3.0	3.8	3.6	9.2	0.7
IgH			6.7	7.1	11.5	8.1	6.2	7.5	4.5	5.3	4.7	4.1	3.5	1.8
Igκ				10.3	10.5	8.9	7.5	6.9	8.2	4.8	5.1	4.2	3.0	2.6
TcRβ					9.8	6.6	6.7	7.6	4.8	4.3	2.0	3.4	2.9	-0.3
TcRα						5.0	6.5	9.9	5.3	6.1	3.5	4.6	2.4	2.1
TcRγ							5.8	8.6	4.2	5.4	3.5	2.1	1.8	0.4
CD8R1								6.1	4.5	5.1	6.0	1.7	5.7	2.0
CD8R2									7.1	5.0	3.5	4.4	3.8	-0.9
CD4										7.7	6.6	1.8	2.8	1.2
PIgRI											6.4	2.6	5.4	-0.3
PIgRIII												4.9	7.0	1.2
OX-2													4.7	0.9
Thy-1														0.6
SOD	-0.3	0.7	-1.8	1.2	0.4	0.1	-0.1	1.8	0.2	-0.9	0.6	0.1	0.6	
RNASE	1.1	0.9	-1.8	2.9	0.6	0.6	-0.6	0.2	0.1	0.4	0.5	1.4	-0.4	
OVALB	0.1	-1.5	-1.7	2.4	-0.8	-0.5	-0.5	0.0	-0.6	0.6	-0.2	-0.5	0.2	

B

	Igλ	Igκ	CH1	CH2	CH3	TcRβ	TcRα	TcRγ	β₂m	MHCI	MHCIIα	MHCIIβ	OX-2	Adeno
		Receptor C domains[c]						C-related sequences[c]						
Igλ		17.6	12.1	8.8	12.5	12.0	4.2	10.4	4.7	9.1	9.6	9.5	4.0	0.5
Igκ			12.8	12.4	13.3	9.8	4.2	7.5	5.7	9.4	7.3	10.6	3.8	0.5
Igγ CH1				8.7	11.4	6.5	3.3	7.5	4.2	8.8	4.9	7.8	1.5	0.9
Igγ CH2					7.3	6.6	1.2	2.9	3.2	2.7	3.2	4.9	3.2	0.3
Igγ CH3						8.5	5.2	6.8	6.4	9.5	5.6	9.1	3.2	1.3
TcRβ							4.3	5.6	4.8	5.7	7.4	6.4	2.8	1.5
TcRα								2.4	0.7	2.3	0.9	2.0	-0.6	0.2
TcRγ									2.0	5.0	6.0	5.3	2.5	-0.3
β₂m										12.0	9.6	12.4	-1.0	0.7
MHCI											8.7	12.6	1.6	1.2
MHCIIα												13.4	-0.4	1.9
MHCIIβ													1.2	1.2
OX-2														0.3

C	Receptor V domains[b]						V-related sequences[b]						
	Igλ	IgH	Igκ	TcR β	TcR α	TcR γ	CD8R1	CD8R2	CD4	PIgRI	PIgRIII	OX-2	Thy-1
IgA	2.4	2.0	1.3	1.0	3.2	1.0	1.5	1.1	2.4	3.9	1.9	2.6	1.1
Igκ	1.9	2.0	1.0	0.6	1.8	2.5	1.8	3.2	2.5	3.2	1.5	2.5	1.2
Igγ CH1	0.3	1.2	0.7	0.8	1.4	1.6	1.2	1.7	2.2	4.6	2.0	1.0	2.8
Igγ CH2	2.3	2.3	2.3	4.1	1.2	3.1	0.7	1.7	2.8	3.8	0.9	2.3	2.8
TcR β	1.7	3.3	1.4	3.3	3.4	1.0	1.4	0.9	2.1	2.3	2.5	3.4	1.9
β2m	-1.7	2.3	1.2	0.5	1.4	-1.1	2.0	1.9	2.2	1.3	1.9	-0.8	2.0
MHCI	0.9	1.6	3.0	3.6	0.6	2.5	1.8	3.8	2.1	1.8	1.2	1.9	2.0
MHCIIα	1.8	2.3	0.9	1.9	1.7	1.3	1.1	2.7	1.2	3.3	2.1	1.8	2.9
MHCIIβ	0.7	4.0	2.0	1.6	3.7	0.8	2.5	3.2	3.6	3.4	1.8	0.4	2.0

D	V or V-related sequences plus N-CAM[b]														
	Igλ	Igκ	IgH	TcR β	TcR α	TcR γ	CD8R1	CD8R2	CD4	PIgRI	PIgRIII	Po	N-CAM2	N-CAM3	N-CAM4
RAT CD2 I	1.8	2.0	0.0	0.8	2.6	2.4	3.2	1.3	2.4	4.4	6.1	4.3	4.7	3.0	3.9
HUM CD2 I	1.6	2.8	0.7	1.6	2.2	3.0	2.0	2.0	4.6	2.7	2.2	4.4	5.7	4.4	3.1

[a]Alignment of Ig V and C-domains and related sequences using the ALIGN program (Dayhoff et al., 1983) with scoring based on the mutation matrix (250 PAMS) with a bias of 2 and a break penalty of 8. In the program the best alignment for two sequences is scored on the basis of the mutation matrix of scores for all possible amino acid substitutions that are derived from the frequency of interchange between amino acids in homologous protein of distantly related species. The sequences were then scrambled and rescored 100 times to give a mean random score and standard deviation and then the score for the real comparison is given as the number of SD units away from the mean random score. (A) Shows alignment scores of V-domains and V-related sequences amongst themselves along with comparisons of adenovirus E3 glycoprotein sequence and control sequences from superoxide dismutase, ribonuclease, and ovalbumin. (B) Scores for comparisons amongst C-domain and C-related sequences. (C) Scores for V- and V-related sequences versus C- and C-related sequences. (D) Scores for human and rat CD2 domain I versus various V-domain, V-ralated, and N-CAM sequences. The sequences used in the alignments were either from the given references or the NBRF protein data base (NBRF code indicated). [Protein Identification Resource (PIR), National Biomedical Research Foundation, Georgetown University, Medical Center, Washington, D.C.]. The numbers refer to the region of sequence used in the alignment. The following V-domain and V-related sequences are used: IgA = Ig lambda-1 precursor, Mouse MOPC 104E (20–129) [L1MS4E]; IgH = Ig heavy V-II, Human Newm (1–115) [G1HUNM]; IGκ = Ig kappa V-I, Human Roy (1–108) [K1HURY]; TcR β = T-cell receptor beta precursor, Human YT35 (20–135) [RWHUVY]; TcR α = T-cell receptor alpha precursor, Mouse (20–132) [RWMSAV]; TcR γ = T-cell receptor gamma precursor, Mouse (19–131) [RWMSV1]; CD8R1 = CD8 chain 1, rat (1–113) (Johnson et al., 1985); CD8R2 = CD8 chain 2, rat (1–109) (Johnson and Williams, 1986); CD4 = CD4, Human (1–110) (Maddon et al., 1985); PIgRI = Poly-Ig receptor, Rabbit (24–134) [QRRBG]; PIgRIII = Poly-Ig receptor, Rabbit (240–344) [QRRBG]; OX-2 = OX-2 antigen rat (1–111) (Clark et al., 1985); Thy-1 = Thy-1 precursor, Rat (20–124) [TDRT]; Adeno = Adenoviral E3 glycoprotein (19–119) [Q6ADE]; SOD = Superoxide dismutase, Bovine (28–138) [DSBOCZ]; RNASE-= Pancreatic Ribonuclease, Bovine (1–110) [NRBO]; OVALB = Ovalbumin, Chicken (1–110) [OACH]. The following C domains or C-related sequences are used; in the data base receptor C-domains begin at sequence 1. IgA = Ig lambda, Human (1–105) [L2HU]; Igκ = Ig kappa, Human (1–106) [K3HU]; Igγ CH1 = Ig gamma-1, Human (1–103) [GHHU]; Igγ CH2 = Ig gamma-1, Human (118–223) [GHHU]; Igγ CH3 = Ig gamma-1, Human (223–326) [GHHU]; TcR β = T-cell receptor beta, Human YT35 (4–114) [RWHUCY]; TcR α = T-cell receptor alpha C region, Mouse (1–94) [RWMSAC]; TcR γ = T-cell receptor non-alpha, Mouse (8–1080 [RWMSC1]; β2M = Beta-2-microglobulin precursor, Human (22–119) [MGHUB2]; MHC I = MHC class I alpha precursor, Human (201–298) [HLHU12]; MHC II α = Class II DR alpha precursor, Human (105–207) [HLHUDA]; MHC II β = Class II DR beta (clone pII-beta-4), Human (92–192) [HLHU4D]; OX-2 = OX-2 antigen, Rat (105–202) (Clark et al., 1985). In the comparisons between CD2 and other domains the CD2 sequences used were as in Fig. 5 and the other sequences started 16 residues before and 20 residues after the 2 Cys residues that form the conserved disulphide bond.

above 2 S.D. units were obtained. Furthermore, the conserved sequence patterns that are notable around the Cys residues of the Ig domains and the domain segments shown in Fig. 1 could not be seen by eye in the E3 sequence.

The analyses in Table IIA–C were done before the sequences for N-CAM, CD3-ϵ, 1B236, PDGF receptor, CSF-1 receptor, α_1B-glycoprotein, and macrophage Fc receptor were published. These sequences are all likely to have folding patterns similar to C domains, but in addition have some striking similarities to V domains in their conserved patterns of sequence. The statistical argument that these sequences are Ig-related is as strong as the other sequences in Table IIA–C, and this is also the case for the domains from the Po protein of peripheral myelin and the proteoglycan link protein, which both have typical V-domain patterns.

Amongst all the new domains claimed to be Ig-related, the most controversial is probably the claim for domains that are Ig-related yet lack the conserved disulphide bond (Fig. 1). Statistical analysis for domain I of CD2 is shown in Table IID with comparisons mainly against V or V-related domains. The results provide a convincing case for the relationship.

4. Categories of Molecules in the Ig Superfamily

4.1. Antigen Receptors and Associated Molecules

The antigen combining site of Igs and TcRs consists of a heterodimeric association of V domains. These structures are extensively discussed elsewhere in this volume. One significant feature of the TcR:antigen interaction is that this involves corecognition of the MHC antigens and thus provides one example of interaction between structures that are themselves related in evolution.

CD3 Antigens

These antigens were identified on human lymphocytes with MAbs (Acuto *et al.*, 1983) and were found to be associated with the T-cell receptor heterodimer. Three CD3 chains have been sequenced and these are called the γ, δ, and ϵ chains (Krissansen *et al.*, 1986; van den Elsen *et al.*, 1984; Gold *et al.*, 1986). Antibodies against the CD3 chains can trigger division of T lymphocytes and it is believed that the CD3 chains are involved in signal transduction after the TcR structure binds antigen (van Wauwe *et al.*, 1980; Samelson *et al.*, 1985).

In the structures of the CD3 chains the features initially noted were the cytoplasmic domains, which were considerably larger than those for the TcR α and β chains, and the presence of a charged residue in each putative transmembrane sequence. It was suggested that these together with charged residues in the TcR transmembrane chain may be involved in signal transduction (Krissansen *et al.*, 1986; van den Elsen *et al.*, 1984; Gold *et al.*, 1986). All the CD3 chains have a single Ig-related domain in their extracellular sequence (Gold *et al.*, 1987a) but this was not obvious at first because these domains are truncated in the portion between the putative disulphide bonds. However a clear relationship to the Ig superfamily can be seen for human CD3-ε chain and through this the relationships of CD3-γ and δ was established (Gold *et al.*, 1987b). The best set of scores with CD3 chains in the ALIGN program are obtained with N-CAM domains, and this seems particularly interesting since the genetic loci for the CD3 chains (Gold *et al.*, 1987a; and G.W. Krissansen and M.J. Crumpton, personal communication), N-CAM and Thy-1 (Nguyen *et al.*, 1986), are all located on the q23 band of human chromosome 11.

4.2. MHC Antigens

The chain structures for class I and class II antigens are shown in Fig. 1 and these have been discussed extensively elsewhere (Lew *et al.*, 1986; Kaufman *et al.*, 1984). In each chain of these antigens there is one segment that is clearly related to Ig C domains and these segments are believed to be positioned adjacent to the lipid bilayer as shown in Fig. 1. The β_2-m structure has been shown by X-ray crystallography to have a fold that is almost identical to that of an Ig C_H3 domain (Becker and Reeke, 1985).

4.3. Antigens Associated with β_2 Microglobulin

The mouse TL and Qa antigens were discovered by immunizing between mouse strains with lymphoid cells (Boyse and Old, 1969; Flaherty, 1976). From the outset these antigens were of particular interest because they were found to be genetically linked to the MHC antigens (reviewed in Lew *et al.*, 1986; Vittetta and Capra, 1978). TL is found on thymic leukemias and, in some strains of mice, on normal thymocytes and activated T lymphocytes, but is not known to be expressed in other tissues. Qa antigens include a number of different structures and show a broader tissue distribution. TL and Qa antigens are associated with β_2-m (Lew *et al.*, 1986; Vittetta and Capra, 1978) and sequences of the H chains from these antigens are very similar to those of class I MHC antigens (Fisher *et a.*, 1985; Devlin *et al.*, 1985). The functions for TL and Qa antigens are unknown, but they may act as a ''library'' of genes with potential for adoption in evolution as class I restriction elements, perhaps by a process of gene conversion (Steinmetz, 1984).

The human CD1 antigens (NA1/34, T6) were identified with mouse MAbs and a set of structures with different H chains is now known to be included in this group (McMichael *et al.*, 1979; Bernard *et al.*, 1984; Amiot *et al.*, 1986).

These H chains are associated with β_2-m but the interaction seems to be of lower affinity than that seen between the H chains of TL or Qa antigens and mouse β_2-m (Amiot *et al.*, 1986). The sequence of one CD1 H chain is known to be class I-like but with a much lower level of identity than seen between TL or Qa H chains and mouse class I H chain (Calabi and Milstein, 1986). This CD1 structure is not genetically linked to the human MHC, and the CD1 antigens should not be thought of as being equivalent to mouse TL or Qa. Rather CD1 constitutes a separate category of β_2-m binding structures that appear to have diverged in evolution from both class I and II MHC antigens to about the same extent that class I and II have diverged from each other (Calabi and Milstein, 1986).

4.4. Receptors for Ig

4.4.1. Poly Ig Receptor

IgA secreted into the gut or at other surfaces has a dimeric structure and includes an additional polypeptide called secretory component (SC) (reviewed in Solari and Kraehenbuhl, 1985). The SC in this complex is the product of proteolytic cleavage of a receptor that binds dimeric IgA at the basal surface of an epithelial cell for transport to the luminal surface. The intact receptor is named Poly Ig R because both multimeric IgA and IgM can be transported (Mostov *et al.*, 1984).

The sequences of rabbit Poly Ig R (Mostov *et al.*, 1984) and human SC (Eiffert *et al.*, 1984) both show segments related to Ig V domains and other V-like structures as shown in Fig. 1. Four of the poly Ig R segments are clearly V-like but the fifth is less clear and has some similarities to C domains.

4.4.2. Mouse Macrophage IgG Receptor

This molecule was first identified via its function of binding aggregated mouse IgG_1 or IgG_{2b}. Then a MAb was produced with which the Fc receptor was purified (Unkeless, 1979), leading to protein sequence and cloning via oligonucleotide probes (Lewis *et al.*, 1986; Ravetch *et al.*, 1986). The structure has two Ig-related domains as shown in Fig. 1, which probably are folded like C domains even though they show some of the conserved patterns of V domains. Domain I of the FcR matches well with the Ig-like domains of MHC class II antigens. For example ALIGN scores of 6.9, 7.2, and 6.2 were obtained for Ig-like domains from β chains of human DC, DR, and SB antigens respectively (unpublished data). Other impressive matches are seen with N-CAM, with which both domains of the FcR sequence can be matched. For example the two domains of FcR can be matched with domains 3 and 4 of N-CAM to give 46 identities out of 162 aligned residues and an ALIGN score of 7.6 SD.

A key point of functional interest with the FcR is that it provides another case of recognition between molecules that are both within the Ig superfamily.

4.5. T-Lymphocyte Marker Antigens

4.5.1. CD2 Antigen

This antigen has been defined with MAbs in human (Howard *et al.*, 1981) and rat (Williams *et al.*, 1987) and is found on most thymocytes, all mature T cells, NK cells, and, in the rat, on spleen macrophages. MAbs against human CD2 can be mitogenic for T lymphocytes (Meuer *et al.*, 1984) and can also interfere with T-lymphocyte adhesion reactions (Shaw *et al.*, 1986). Human CD2 is the molecule that mediates rosetting between human T cells and sheep erythrocytes (Howard *et al.*, 1981).

The CD2 sequence shows a highly significant relationship to the second half of the CD4 antigen sequence (Fig. 1) (Williams *et al.*, 1987; Sewell *et al.*, 1986). There are two extracellular CD2 domains and both have similarities to Ig superfamily sequences without either having the typical conserved characteristics. CD2 domain I scores well with CD4, Poly Ig R, and N-CAM Ig-related sequences (Table IID). The intracellular domain of 116 residues is notable for its very high content of proline residues (Williams *et al.*, 1987; Sewell *et al.*, 1986).

4.5.2. CD4 Antigen

This antigen has been identified with MAbs in rat (called W3/25) (Williams *et al.*, 1977), human (OKT4) (Reinherz *et al.*, 1979), mouse (L3T4) (Dialynas *et al.*, 1983), and other species (Classon *et al.*, 1986a) and is found on thymocytes and a subset of T lymphocytes including T helper cells in all species (Mason *et al.*, 1983). In some species it is also found on macrophages (Jefferies *et al.*, 1985). Antibodies against CD4 almost totally inhibit mixed lymphocyte responses suggesting an important role for this antigen in the activation of T helper cells (Webb *et al.*, 1979).

Human and mouse T cells that expressed CD4 were found to include cytotoxic cells whose specificity was associated with class II MHC antigens and, on this basis, it was suggested that CD4 may bind to class II MHC to control the activation of CD4$^+$ T cells (Swain, 1983; Reinherz *et al.*, 1983). The evidence on this point seems contradictory (Gay *et al.*, 1986; Bank and Chess, 1985) and at this stage the exact role of CD4 is unclear. Antibodies to human CD4 prevent *in vitro* infection of human T cells with AIDS virus (Dalgleish *et al.*, 1984).

CD4 has been sequenced in human (Maddon *et al.*, 1985), mouse (Tourvieille *et al.*, 1986), and rat (Clark *et al.*, 1987) to reveal a structure with four extracellular domains with disulphide bonds as shown in Fig. 1 (Classon *et al.*, 1986b). The NH$_2$-terminal part of CD4 has a typical V-like domain (Maddon *et al.*, 1985) while the other three domains are atypical with domains II and IV being truncated and domain III lacking a disulphide bond. The similarities of domains II and III to Ig-related sequences was shown by comparison with the Poly Ig R sequence (Clark *et al.*, 1987).

4.5.3. CD8 Antigen

CD8 antigen was identified in the mouse when the Lyt-2 and Lyt-3 alloantigens were described (Boyse and Old, 1969). Subsequently it was found that the two sets of determinants were on separate but disulphide-linked chains (Ledbetter *et al.*, 1981) and that the genetic loci for these chains were closely linked to each other and to Ig κ chains (Gottlieb, 1974). The Lyt-2 and -3 antigens were found on most thymocytes but amongst mature T lymphocytes were believed to be restricted to cytotoxic and suppressor T cells and their precursors (Cantor and Boyse, 1975).

In other species the CD8 antigen was identified with monoclonal antibodies and the general pattern of distribution is as for the mouse with the additional finding that rat and human NK cells can also express CD8 (reviewed in Johnson *et al.*, 1985). In mouse and humans it has been shown that CD8+ T cells are restricted by class I but not class II MHC antigens and the function of these cytotoxic cells can be inhibited by anti-CD8 antibodies (MacDonald *et al.*, 1982). These findings led to the suggestion that CD8 and class I antigens interact directly in the triggering of a CD8+ T cell (Swain, 1983; Reinherz *et al.*, 1983). However there is no clear-cut evidence for this.

The structures of human, mouse, and rat CD8 antigens have been established. One surprising feature is that the human structure has been suggested to be a homodimer while the rodent structures are heterodimers (reviewed in Johnson *et al.*, 1985). Chain 1 as shown in Fig. 1 was initially cloned for human (Littman *et al.*, 1985; Sukhatme *et al.*, 1985a) and then mouse (Nakauchi *et al.*, 1985) and rat (Johnson *et al.*, 1985) while chain 2 has been recently cloned in the rat (Johnson and Williams, 1986). Now a gene and mRNA coding for chain 2 have been identified in humans (P. Johnson, unpublished) and it remains a possibility that human CD8 is a heterodimer with the second chain having gone undetected in immunochemical studies. The CD8 heterodimer appears to have a particularly close relationship to Ig or TcR V domains and in particular the V-like domain in Chain 2 gives high ALIGN scores with receptor V domains and has a segment of sequence that closely matches some receptor J pieces (Johnson and Williams, 1986) (Fig. 6).

4.6. Brain/Thymus Antigens

4.6.1. Thy-1 Antigen (Reviewed in Williams and Gagnon, 1982)

Thy-1 (theta) was discovered when the mouse Thy-1.1 and Thy-1.2 allotypes were identified (Reif and Allen, 1964). Anti-Thy-1.1 antiserum cross reacted with rat Thy-1 leading to its purification and the identification of Thy-1 in human and other species. In mice, Thy-1 is expressed in large amounts on thymocytes and in the brain and also on T lymphocytes and fibroblasts. It was

```
RAT CD8 37 Kd Chain                    V  V  F  G  T  G  T  K  L  T  V  V  D
HUMAN Jλ VIL                           V  V  F  G  G  G  T  K  L  T  V  L  G
HUMAN Jλ MOT                           V  V  F  G  T  G  T  M  V  T  V  L  G
MOUSE Jλ CONSENSUS                     W  V  F  G  G  G  T  K  L  T  V  L  G
MOUSE Jκ CONSENSUS                     W  T  F  G  G  G  T  K  L  E  I  K  R  A
ALL J_H CONSENSUS        F  F  D  V  W  G  Q  G  T  T  V  T  V  S  S
MOUSE JTβ 7           C  S  Y  E  Q  Y  F  G  P  G  T  R  L  T  V  L
MOUSE JTα TA39       G  T  Y  Q  R  F  G  T  G  T  K  L  Q  V  V  P  N
HUMAN MRC OX-2             N  T  F  G  F  G  -  K  I  S  G  T  A
```

Figure 6. Comparisons of sequences from rat CD8 chain 2 and MRC OX-2 antigen with J piece sequences from Ig or Tcr V domains as given in Johnson and Williams (1986). Residues in CD8 chain 2 and OX-2 that are identical to J piece sequences are boxed. The dash indicates a gap inserted to improve the alignments of the OX-2 sequence. The CD8 and OX-2 sequences begin 8 and 3 residues respectively after the COOH-terminal Cys of the putative conserved disulphide bond. Thus the CD8 sequence is in position almost exactly equivalent to the J piece sequences.

the first antigen used to separate T and B cells in functional studies. In other species Thy-1 expression is conserved in the brain, where the molecule is a major surface component of neurons, and on fibroblasts, but expression varies greatly in the lymphoid cells. Rat T cells lack Thy-1 even though thymocytes express 10^6 molecules per cell and in humans the antigen is expressed on few thymocytes or T cells. Given the variation of expression on lymphoid cells, it is curious that a powerful mitogenic stimulus can be given to mouse T cells by cross-linking antibodies bound to cell surface Thy-1 and culturing with phorbol myristic acid (Kroczek *et al.*, 1986).

Rat and mouse Thy-1 have sequences of 111 and 112 amino acids, respectively, with two interchain disulphide bonds, 3 N-linked carbohydrate structures, and a sequence that looks like an Ig-V domain (Williams and Gagnon, 1982). Human sequence determined at the cDNA level is 68% identical to the rat sequence (Seki *et al.*, 1985). Thy-1 has the unusual feature of having a glycophospholipid tail attached at the COOH-terminus for membrane integration (Tse *et al.*, 1985). From the cDNA coding for Thy-1 an extra sequence of 31 residues is predicted in addition to the sequence determined at the protein level (Seki *et al.*, 1985). This sequence is believed to function as a signal for the biosynthetic machinery that cleaves the protein after residue 111 and attaches the glycophospholipid tail (Conzelmann *et al.*, 1986).

4.6.2. MRC OX-2 Antigen

This antigen was identified with a mouse MAb raised against rat thymocyte glycoproteins (Barclay and Ward, 1982). The OX-2 antigen is found on thymocytes, B lymphocytes, and follicular dendritic cells, most if not all endothelium, neurons, and smooth muscle (Webb and Barclay, 1982). It, thus, has a diverse pattern of expression as is the case for Thy-1. These two antigens also appear similar in that they both show differences in the glycosylation of their brain and thymus forms (Barclay and Ward, 1982). The function of OX-2 gly-coprotein is unknown.

OX-2 was sequenced to show a structure similar to a single TcR chain or Ig L chain (Clark *et al.*, 1985) (Fig. 1). Human OX-2 has been identified at the genomic DNA level and the sequence is 78% identical overall to rat OX-2 (McCaughan *et al.*, 1987a).

4.7. Neural Cell-Surface Molecules

4.7.1. N-CAM

These glycoproteins were identified as the molecules responsible for the specific formation of aggregates by chick embryonic neural cells (Edelman, 1985). This interaction is thought to be due to homophilic recognition between N-CAM molecules on different cells (Rutishauser, *et al.*, 1982). MAbs were produced against N-CAM allowing the simple purification of chicken N-CAM (Edelman, 1985) and the identification of mouse N-CAM (He *et al.*, 1986). The N-CAM molecule seems restricted to neurons and glial cells in adults, but is found on a wide range of embryonic cells types (Edelman, 1985).

Chicken N-CAM was partially sequenced from cDNA clones obtained using oligonucleotide probes and found to contain four Ig-related domains with the possibility of further domains to be identified when the NH_2-terminus is completed (Hemperley *et al.*, 1986). Thus the structure looks rather like the Poly Ig R in terms of domain organization. However, it differs from the Poly Ig R in that the N-CAM domains are likely to have C-domain folds even though some of the conserved patterns of V domains are seen.

The N-CAM molecule shows interesting heterogeneity in its carbohydrate structures and cytoplasmic domains. In the embryo, but not the adult, the molecule expresses carbohydrates with polysialic acid chains attached (Edelman, 1985) and this is thought to impede N-CAM interactions. In the cytoplasmic part of the molecule three alternative molecular forms are found and are believed to be produced by alternative splicing of mRNA (Murray *et al.*, 1986). Two forms include a transmembrane sequence with cytoplasmic domains of different sizes while the third form has a glycophospholipid domain similar to that found on Thy-1 for membrane attachment (He *et al.*, 1986). The similarities to Thy-1 include neuronal expression, the glycophospholipid tail, and linkage in the genome with both genes being found on band q23 of chromosome 11 along with CD3 (Nguyen *et al.*, 1986).

4.7.2. 1B236 Brain Molecule

This molecule was identified as a rat-brain-specific cDNA clone coding for a structure with some of the characteristics expected for a precursor of neurotransmitter peptides (Sutcliffe *et al.*, 1983). However, in further studies in which the sequence was completed, five segments showing Ig-like domains were identified followed by a transmembrane protein sequence and cytoplasmic domain predicted to have 90 amino acids (Lai *et al.*, 1987). Thus it seems likely that the molecule will have the structure indicated in Fig. 1, which shows considerable similarity in organization to the Poly Ig receptor. However the molecule has truncated domains as does N-CAM and in ALIGN comparisons shows strong relationships to the N-CAM domains (Lai *et al.*, 1987).

The translated product of the 1B236 cDNA has been identified in immunohistochemistry using antibodies raised against peptides from the part of the sequence predicted to be in the cytoplasmic domain. It appears that this structure is restricted to the brain and that it acts as a marker for neuron subsets in the adult rat (Lai *et al.*, 1987).

4.7.3. Po Glycoprotein of Peripheral Myelin

The Po glycoprotein constitutes about 50% of the protein in the multilayered membrane produced by Schwann cells to form the peripheral myelin sheath (Greenfield *et al.*, 1973). Po is absent from myelin of the central nervous system. Po was sequenced at the cDNA level to suggest a transmembrane glycoprotein with 124 extracellular and 69 intracellular amino acids (Lemke and Axel, 1985). It was suggested that the extracellular parts of Po undergo a homophilic interaction when the myelin membrane compacts as shown in Fig. 7. The relationship of Po to Ig was noted when the 1B236 sequence was analyzed (Lai *et al.*, 1987) and the external 124 residues fit exactly with one Ig V-domain-type fold. ALIGN scores of 5.1, 7.1, 7.2, and 7.2 SD units are obtained in comparisons with the V or V-like domains: V_H Newm, CD4 domain I, Poly Ig R domain III, and MRC OX-2 domain I.

4.8. Growth Factor Receptors

Recently, sequences have been determined for the receptor for platelet-derived growth factor (PDGF) (Yarden *et al.*, 1986) and for c-*fms* (Coussens *et al.*, 1986) which is the cellular equivalent of the v-*fms* oncogene and is thought to be the receptor for macrophage colony-stimulating factor (Scherr *et al.*, 1985) (henceforth called CSF-1 receptor). These receptors both have a single putative transmembrane sequence and large cytoplasmic domains that have tyrosine kinase activity. The extracellular domains include 499 residues for the PDGF receptor and 493 residues for CSF-1 receptor. The relationship of these sequences to Ig was noted when the 1B236 sequence was analyzed (Lai *et al.*, 1987) and five Ig-related domains can be detected in the extracellular part of the PDGF and CSF-1 receptors. These sequences thus look rather like the Poly Ig R, α_1B-GP, N-CAM, and 1B236 structures.

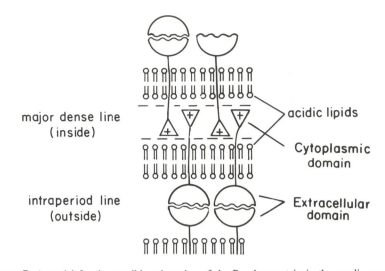

Figure 7. A model for the possible orientation of the Po glycoprotein in the myelin membrane. Semicircles (⌒) represent the Po extracellular domain and triangles (△) the Po cytoplasmic domain. Cytoplasmic domains are illustrated as forming the major dense line through electrostatic interaction with the head groups of acidic lipids present in the cytoplasmic leaflet of the opposed bilayer. Extracellular domains are illustrated as interacting with each other to form the intraperiod line. The figure is reproduced with permission from Lemke and Axel (1985).

Preliminary ALIGN analysis indicates that the first 3 domains from the NH_2-terminus in each sequence are of the N-CAM type with the possibility of a C-type fold but with V-like patterns of sequence around the second cysteine residue. The fourth domain (residues 286–394) in the PDGF receptor is also about the length of N-CAM domains but lacks Cys residues. However, good ALIGN scores are obtained with N-CAM and some V-related domains. For example, comparison of PDGF receptor domain IV with N-CAM domains II and III shows identities at about 25% and ALIGN scores of 5.9 and 5.2 SD units, respectively. The fifth domain of PDGF and CSF-1 receptors is longer than the other four and may have a V-like fold.

4.9. Molecules Not Known to be Cell-Surface Associated

4.9.1. α_1 B-glycoprotein

This is a glycoprotein of unknown function found in serum at the level of 0.22 mg/ml. When sequenced five Ig-like domains were identified and typical conserved disulphide bonds were established for three of these domains (Ishioka *et al.*, 1986). The α_1 B-glycoprotein domains give good scores in the ALIGN program with V and V-related domains (not shown) but assignments of putative β-strands suggest a C-domain fold. Thus the α_1 B-glycoprotein structure looks rather like Poly Ig R, or IB236, or N-CAM in organization without any known structure for attachment to a cell surface.

4.9.2. Basement Membrane Link Protein

In the basement membrane, link proteins exist to stabilize the interaction between a core of hyaluronic acid and the proteoglycan branch structures. The NH_2-terminal 100 residues of the link protein from bovine nasal cartilage has been sequenced and this sequence constitutes a typical V-related domain (Bonnet *et al.*, 1986). Preliminary ALIGN analysis gives scores of 6–7 SD for Vλ (MOPC104E), human CD4, and rabbit Poly Ig R domains I and III.

5. Gene Organization

The gene organization for structures, as published by early 1986, is shown in Fig. 8 and at that time the generalization could be made that, in all cases, the Ig domains or domain-related segments were largely encoded by single exons. The gene organization for 1B236 also shows introns separating each of the five Ig-related domains (Lai *et al.*, 1987). More recently some exceptions have been found. In the rabbit Poly Ig R gene one exon accounts for domains II and III (Deitcher and Mostov, 1986) and there is an intron between nucleotides coding for the Cys residues of the conserved disulphide bond in domain I. This comes after amino acid residue 83, which would be expected to correspond to a position at the base of β-strand E in the Ig-related fold of domain I (see Fig. 3) (K. Mostov, personal communication). Also in domain I of CD4 an intron is found at a central position in the domain (R. Axel, personal communication).

Taken together the genetic data strongly support the hypothesis that the Ig-related sequences evolved by gene duplication of domain-sized genetic units. However the introns within domains raise the question as to whether a smaller structure may have preceded the full domain structure. A primordial half-domain has been previously postulated on the basis of protein sequence (Bourgois, 1975) and structural considerations (McLachlan, 1980).

The V domains of antigen receptors present another variation on the single exon theme in that the final V-domain sequence is assembled by rearrangement to splice a gene segment, encoding most of the V-domain, onto the small J or D-plus-J segments. The J segments in V_L and V_H encode β strand G of the V-domain fold and in V_H the D segment encodes the loop of sequence between β-strands F and G (Fig. 3). In contrast, in chain 2 of CD8 antigen (Johnson and

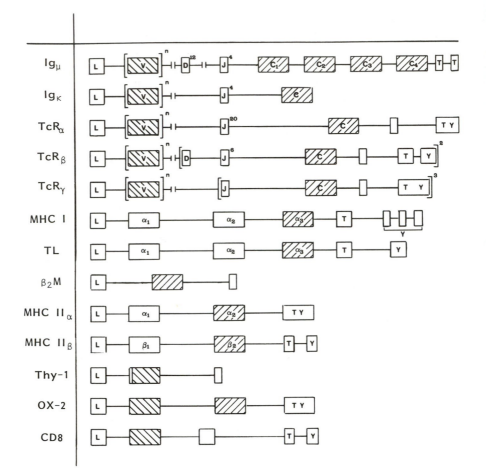

Figure 8. Gene organisation of coding sequences for molecules in the Ig superfamily. V and V like exons are shaded ▨ , C and C like exons are shaded ▨ . Untranslated exons are not shown. L represents leader, T transmembrane, Y, cytoplasmic. Data is derived from the following references: Igμ (Alt *et al.*, 1986), Igκ (Joho *et al.*, 1980), TcR α (Yashikai *et al.*, 1985), TcR β (Gascoigne *et al.*, 1984), TcR γ (Hayday *et al.*, 1985), MHC I (Lew *et al.*, 1985), TL (Fisher *et al.*, 1985), β₂-m (Parnes *et al.*, 1986), MHC II α and β (Trowsdale *et al.*, 1985), Thy-1 (Giguere *et al.*, 1985), OX-2 (McCaughan *et al.*, 1987a), and CD8 (Sukhatme *et al.*, 1985). The Qa genes are not shown as it is not clear whether the published structures (Devlin *et al.*, 1985) represent functional genes. The CD8 structure is derived from known intron/exon boundaries (Sukhatme *et al.*, 1985). The diagram is not drawn to scale.

Williams, 1986) and MRC OX-2 antigen (McCaughan *et al.*, 1987a), the sequences with similarities to J-piece sequences are not separated in the genome from DNA coding for the rest of the V-like domain. It is possible that the gene arrangement seen in the receptor domains evolved by the insertion of intervening DNA into sequences like those seen in chain 2 of CD8 or MRC OX-2 antigen.

Genetic mapping of genes coding for structures in the Ig-superfamily shows that these are usually unlinked. Notable exceptions are the well-known gene clusters coding for the Ig heavy chains and the MHC antigens. Other gene clusters of interest include the loci for CD8 antigens and Vκ chains and the group of genes found on chromosome 11 band q23 that includes N-CAM, Thy-1, and CD3 chains (Table I).

In the case of CD8 and Vκ, it seems unlikely that these molecules shared an immediate precursor distinct from the precursors for other antigen receptor chains since the CD8 genes do not have a structure similar to receptor genes with J pieces on distinct exons. It could be that the Vκ and CD8 genes have remained in a chromosome region where evolution of a set of genes giving rise to heterodimers with immune functions occurred.

In the case of N-CAM, Thy-1, and CD3 there is the possibility that a genetic region that was crucial in the early evolution of the Ig-superfamily is being marked. Thy-1 and the Po myelin protein are the only members of the superfamily that exist as single domains not associated with other chains and these molecules may have been derived with little change from primitive structures. Also the case for N-CAM being a primitive structural type can be made on the basis that the molecule is expressed widely in embryos. CD3 is not an obvious candidate as a primitive form, but it is interesting that CD3 sequences seem to align best with N-CAM sequences (Gold et al., 1987a). CD3 may have derived in a lineage from the single domain type that duplicated to give the N-CAM structure.

6. Evolution of Structural Types in the Ig Superfamily

A key question is whether the Ig superfamily is a divergent set of genes or convergent evolution to a favorable structural type has occurred. The argument against convergent evolution is that it is not obvious that the production of similar structural types should require strong similarities in sequence patterns. The Ig-related domains show enormous variation in sequence even in the stretches that form the β-strands and there are no invariant sequences that seem required for the Ig fold. This point is also made from the structure of superoxide dismutase, which has similarities to the Ig fold without any discernible patterns of sequence similarity (Richardson et al., 1976). The patterns of sequence similarity within the Ig-related structures seem consistent with the idea that a stable domain unit first evolved and then variants of this were produced and diverged in different lines in which various sequence patterns were maintained. It can be argued that no direct selection for sequence occurred in this divergence since the determinants that are involved in possible recognition events would not be those that are conserved between the functionally different molecules. It could be that the conserved features are seen because it was essential that a structure that was stable to proteolytic enzymes be maintained for functions that involved exposure to the extracellular environment. Thus the stable fold would have to be maintained while variation in the loops and on the outer faces was occurring in order to provide different biological specificities. Solutions to the fold that were drastically different to immediate precursors might not be seen because pathways to alternative structures would not maintain a stable domain structure.

Assuming divergent evolution there is the possibility of attempting to draw family trees. However this is very difficult for the Ig superfamily because the sequences show great variation. At this time no attempt has been made to measure evolutionary distances between domain segments as shown in Fig. 1. Many of the sequences are new and analysis is at a preliminary stage. One new feature that can be added to the thinking shown in Fig. 2, is the common occurrence of multiple domain, single chain structures including Poly Ig R, N-CAM, 1B236, PDGF and CSF-1 receptors, CD4, and α_1 B-glycoprotein. It seems possible that, at every stage after the evolution of the first domain, more than one set of each structural type may have evolved and these may have undergone parallel evolution thereafter.

A key point in trying to determine evolutionary pathways will be to find Ig-related molecules in invertebrates. A Thy-1 homologue was sought in invertebrate neural tissue and a structure with interesting similarities to Ig was identified (Williams *et al.*, 1985; Williams and Gagnon, 1982). However the case for this being Ig-related is less strong than for the molecules in Fig. 1 and Table I and further data on invertebrate structures is needed before clear interpretations can be made.

7. Possible Functional Derivation of the Ig Superfamily

In considering functional aspects the following points are relevant:

1. No enzymatic activity is known to be mediated by Ig or Ig-related domains.

2. Most of the known structures are found attached to cell surfaces.

3. The known functions of most molecules include binding to another molecule as a part of, or a trigger for, a subsequent event. For example: Ig, TcR, PDGF receptor, and CSF-1 receptor all act as triggering molecules for turning on cell division and other activities; MHC interacts with TcR as part of T-cell triggering; Poly Ig R and macrophage FcR bind Igs for transport or triggering of macrophage functions; Ig in antigen:antibody complexes interacts with complement to initiate enzymatic activities in the complement components; N-CAM molecules interact to mediate cell adhesion; CD2 mediates cell adhesion; myelin Po protein may mediate membrane adhesion.

4. In some cases an Ig-related molecule may trigger functional effects on other cells but not directly affect the activities of the cell that displays the molecule. This may be the case with class I MHC involvement in target cell killing

where, as far as is known, the class I antigen on the target plays no role in the killing event other than to trigger the response of the killer cell.

5. A single Ig-domain can interact with itself (C_H3 domains of Ig, Ig L chain dimers) or with different single domains ($V_H:V_L$, $C_H1:C_L$, etc.) within a multimeric chain structure. In the case of N-CAM homophilic interactions appear to occur between the same molecule on different cells (Rutishauser *et al.*, 1982) and homophilic reactions are also suggested for Po proteins between opposed sheets of myelin membrane (Lemke and Axel, 1985). In the case of Po a homophilic reaction could involve only the V-like extracellular segment but in the case of N-CAM it is not yet known whether the adhesion is mediated by the Ig-related part of the molecule (see Fig. 1).

6. Interactions can occur between different members of the Ig superfamily. This is seen in the interactions between TcR and class I or II MHC in the core-cognition of foreign antigen and also in the interaction between Poly Ig R and polymeric Igs and macrophage FcR and IgG. Interactions between CD8 and class I MHC and CD4 and class II MHC have been postulated but these are not established. It seems that interactions within the superfamily may be the rule rather than an exceptional occurrence.

In attempts to propose functional origins for the Ig superfamily, the above points have been taken into account and in particular roles for a single domain have been considered (Williams *et al.*, 1985; Williams, 1982). These ideas are summarized in Fig. 9 where the Ig-related molecules are suggested to have first evolved to mediate recognition between cells that were not involved in immunity. If a moderately large family of recognition structures evolved, then a subset of this family may have been adopted for immune functions at the time of vertebrate evolution. An initial role for Ig-related structures in basic cell recognition events seems to be strongly supported by the findings that N-CAM, the growth factor receptors, and the Po myelin protein are Ig-related.

In Fig. 9 it is also suggested that the Ig-related molecules may have first evolved in neural tissues (Williams *et al.*, 1985). This was first suggested because Thy-1 was a single-domain structure and was conserved between species in neurons. The finding that Po protein also displays a single extracellular V domain strongly reinforces this thinking and it is striking that both neurons and glial cells feature single Ig-related V-like structures as major cell-surface components. The model for Po myelin protein function shown in Fig. 7, which was suggested without the knowledge that Po is Ig-related, corresponds closely to the type of interaction previously postulated as a possible function for a primitive Ig domain (Williams *et al.*, 1985; Williams, 1982). Neural tissues also express the OX-2 antigen, which is a simple two domain structure and the N-CAM and 1B236 structures. N-CAM would seem to be a primitive structure on the basis of its widespread expression in the early embryo and it seems clear that the Ig-related molecules must have been extensively involved in the early evolution of sensory systems. If the Ig-related molecules first evolved for basic cell interac-

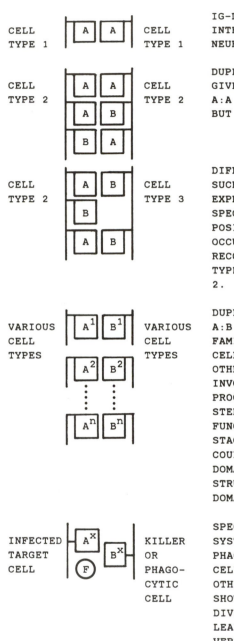

CELL
TYPE 1 | A A | CELL
TYPE 1

IG-DOMAIN-LIKE STRUCTURE FOR
INTERACTIONS BETWEEN PRIMITIVE
NEURAL-TYPE CELLS.

↓

CELL
TYPE 2 | A A / A B / B A | CELL
TYPE 2

DUPLICATION AND DIVERGENCE TO
GIVE A AND B DOMAINS SUCH THAT
A:A AND A:B INTERACTIONS OCCUR
BUT NOT B:B.

↓

CELL
TYPE 2 | A B / B / A B | CELL
TYPE 3

DIFFERENTIAL GENE EXPRESSION
SUCH THAT CELL TYPE 3
EXPRESSES ONLY DOMAIN B.
SPECIFIC CONTROL OF CELL
POSITION OR BEHAVIOUR CAN
OCCUR SINCE CELL TYPE 2 CAN
RECOGNIZE ITSELF OR TYPE 3 BUT
TYPE 3 CAN ONLY RECOGNIZE TYPE
2.

↓

VARIOUS
CELL
TYPES | A^1 B^1 / A^2 B^2 / A^n B^n | VARIOUS
CELL
TYPES

DUPLICATION AND DIVERGENCE OF
A:B SYSTEM TO PRODUCE A LARGE
FAMILY OF MOLECULES FOR
CELL:CELL RECOGNITION AND
OTHER RECEPTOR FUNCTIONS.
INVOLVEMENT IN SPECIFICITY OF
PROGRAMMED CELL DEATH MAY BE A
STEP LEADING TO IMMUNE
FUNCTIONS.   NOTE; BY THIS
STAGE THE A^1-A^n, B^1-B^n UNITS
COULD REPRESENT MULTIPLE
DOMAINS AND TWO CHAIN
STRUCTURES AS WELL AS SINGLE
DOMAINS.

↓

INFECTED
TARGET
CELL | A^x B^x / F | KILLER
OR
PHAGO-
CYTIC
CELL

SPECIFICITY OF CELL-DEATH
SYSTEM CHANGED TO INCORPORATE
PHAGOCYTOSIS OR KILLING OF
CELLS INFECTED WITH VIRUS OR
OTHER PATHOGEN (DETERMINANT
SHOWN AS F), EXTENSIVE
DIVERSIFICATION OF THIS SYSTEM
LEADS TO THE IG-RELATED
VERTEBRATE IMMUNE SYSTEM.

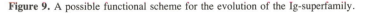

Figure 9. A possible functional scheme for the evolution of the Ig-superfamily.

tions, then a role in immunity would require that some recognition function controlling the behavior of cells within an organism would need to be turned outwards to recognize foreign material. One possibility would be the modification of a recognition system involved in programmed cell death that occurs in the course of cell differentiation. Cell death has been studied during development in *Caenorhabditis elegans* and usually involves death of a cell followed by phagocytosis (Horvitz *et al.*, 1982; Hedgecock *et al.*, 1983). However there are cases where one cell is killed by another prior to phagocytosis, and such a function could provide the basis for immunity if the specificity of the interaction was modified to involve determinants of a common pathogen (Fig. 9). It is interesting that programmed death occurs extensively in the development of neural tissue (Horvitz *et al.*, 1982) since this provides a functional candidate for origin of immunity in the same tissue that has been suggested as a site for the evolution of the Ig-related molecules on the basis of molecular data.

Many more polypeptides have recently been shown to contain Ig-related sequences and are reviewed in Williams and Barclay (1988).

ACKNOWLEDGMENTS. P.J. is a Beit Fellow; G.W.McC. is a C.J. Martin NH and MRC Fellow; we thank the Protein Identification Resource for providing the NBRF data base and ALIGN programmes, Mrs. Caroline Griffin for help with the manuscript and Drs. R. Axel, G.W. Krissansen, R. Milner, K.E. Mostov and M.J. Crumpton for communicating unpublished data.

References

Acuto, O., Hussey, R.E., Fitzgerald, K.A., Protentis, J.P., Meuer, S.C., Schlossman, S.F., and Reinherz, E.L., 1983, The human T cell receptor: appearance in ontogeny and biochemical relationship of α and β subunits on IL-2 dependent clones and T cell tumors, *Cell* **34**:717–726.

Alt, F.W., Blackwell, T.K., Depinho, R.A., Reth, M.G., and Yancopoulos, G.D., 1986, Regulation of genome rearrangement events during lymphocyte differentiation, *Immunol. Rev.* **89**:5–80.

Amiot, M., Bernard, A., Raynal, B., Knapp, W., Deschildre, C., and Boumsell, L., 1986, Heterogeneity of the first cluster of differentiation: characterisation and epitopic mapping of three CDI molecules on normal human thymus cells, *J. Immunol.* **136**:1752–1758.

Amzel, L.M., and Poljak, R.J., 1979, Three-dimensional structure of immunoglobulins, *Ann. Rev. Biochem.* **48**:961–997.

Bank, I., and Chess, L., 1985, Perturbation of the T4 molecule transmits a negative signal to T cells, *J. Exp. Med.* **162**:1294–1303.

Barclay, A.N., and Ward, H.A., 1982, Purification and chemical characterisation of membrane glycoproteins from rat thymocytes and brain, recognized by monoclonal antibody MRC OX-2,

Eur. J. Biochem. **129**:447–458.

Becker, J.W., and Reeke, G.N., Jr., 1985, Three-dimensional structure of β_2-microglobulin, *Proc. Natl. Acad. Sci. USA* **82**:4225-4229.

Bernard, A., Boumsell, L., Dausset, J., Milstein, C., and Schlossman, S.F., eds., 1984, Leucocyte Typing. Human Leucocyte Differentiation Antigens Detected by Monoclonal Antibodies, Springer-Verlag, Berlin.

Bonnet, F., Perin, J.-P., Lorenzo, F., Jolles, J., and Jolles, P., 1986. An unexpected sequence homology between link proteins of the proteoglycan complex and immunoglobulin-like proteins, *Biochim. Biophys. Acta* **873**:152–155.

Bourgois, A., 1975, Evidence for an ancestral immunoglobulin gene coding for half a domain, *Immunochemistry* **12**:873–876.

Boyse, E.A., and Old, L.J., 1969, Some aspects of normal and abnormal cell surface genetics, *Annu. Rev. Genet.* **3**:269–290.

Calabi, F., and Milstein, C., 1986, A novel family of human major histocompatiblity complex-related genes not mapping to chromosome 6, *Nature* **323**:540–543.

Cantor, H., and Boyse, E.A., 1975. Functional subclasses of T lymphocytes bearing different Ly antigens. I. The generation of functionally distinct T-cell subclasses is a differentiative process independent of antigen, *J. Exp. Med.* **141**:1376–1389.

Chatterjee, D., and Maizel, J.V., 1984, Homology of adenoviral E3 glycoprotein with the HLA-DR heavy chain, *Proc. Natl. Acad. Sci. USA* **81**:6039–6043.

Chien, Y., Becker, D.M., Lindsten, T., Okamura, M., Cohen, D.I., and Davis, M.M., 1984, A third type of murine T-cell receptor gene, *Nature* **312**:31–35.

Clark, S.J., Jefferies W.A., Barclay, A.N., Gagnon, J., and Williams, A.F., 1987, Peptide and nucleotide sequences of rat CD4 (W3/25) antigen: evidence for derivation from a structure with four immunoglobulin-related domains, *Proc. Natl. Acad. Sci. USA,* **84**:1649–1653.

Clark, M.J., Gagnon, J., Williams, A.F., and Barclay, A.N., 1985, MRC OX-2 antigen: a lymphoid/neuronal membrane glycoprotein with a structure like a single immunoglobulin light chain, *EMBO J.* **4**:113–118.

Classon, B.J., Tsagarotos, J., Kirszbaum, L., Maddox, J., Mackay, C.R., Brandon, M., McKenzie, I.F.C., and Walker, I.D., 1986a, The L3T4 antigen in mouse and the sheep equivalent are immunoglobulin-like, *Immunogenetics* **23**:129–132.

Classon, B.J., Tsagaratos, J., McKenzie, I.F.C., and Walker, I.D., 1986b, Partial primary structure of the T4 antigens of mouse and sheep: assignment of intrachain disulfide bonds, *Proc. Natl. Acad. Sci. USA* **83**:4499–4503.

Conzelmann, A., Spiazzi, A., Hyman, R., and Bron, C., 1986, Anchoring of membrane proteins via phosphatidylinositol is deficient in two classes of Thy-1 negative mutant lymphoma cells, *EMBO J.* **5**:3291–3296.

Coussens, L., Beveren, C.V., Smith, D., Chen, E., Mitchell, R.L., Isacke, C.M., Verma, I.M., and Ullrich, A., 1986, Structural alteration of viral homologue of receptor proto-oncogene *fms* at carboxyl terminus, *Nature* **320**:277–280.

Dalgleish, A.G., Beverley, P.C.L., Clapham, P.D., Crawford, D.H., Greaves, M.F., and Weiss, R.A., 1984, The CD4 (T4) antigen is an essential component of the receptor for the AIDS retrovirus, *Nature* **312**:763–767.

Dayhoff, M.O., Barker, W.C., and Hunt, L.T., 1983. Establishing homologies in protein sequences, *Meths. Enzymol.* **91**:524–545.

Deitcher, D.L., and Mostov, K.E., 1986, Alternative splicing of rabbit polymeric immunoglobulin receptor, *Molec. Cell. Biol.* **6**:2712–2715.

Devlin, J.J., Weiss, E.H., Paulson, M., and Flavell, R.A., 1985, Duplicated gene pairs and alleles of class I genes in the Qa 2 region of the murine major histocompatibility complex: a comparison, *EMBO J.* **4**:3203–3207.

Dialynas, D.P., Wilde, D.B., Marrack, P., Pierres, A., Wall, K.A., Havran, W., Otten, G.,

Loken, M.R., Pierres, M., Kappler, J., and Fitch, F.W., 1983, Characterization of the murine antigenic determinant, designated L3T4a, recognized by monoclonal antibody G.K. 1.5: Expression of L3T4a by functional T cell clones appears to correlate primarily with class II MHC antigen reactivity, *Immunol. Rev.* **74:**29–56.

Edelman, G.M., 1985, Cell adhesion and the molecular processes of morphogenesis, *Ann. Rev. Biochem.* **54:**135–169.

Edelman, G.M., 1970, The covalent structure of a human γG-immunoglobulin. XI. Functional implications, *Biochemistry* **9:**3197–3205.

Eiffert, H., Quentin, E., Decker, J., Hillemeir, S., Hufschmidt, M., Klingmuller, D., Weber, M.H., and Hilschmann, N., 1984, Die primarstruktur der menschlichen freien sekretkomponente und die anordnung der disulfidbrucken, *Hoppe-Seyler's Z. Physiol. Chem.* **365:**1489–1495.

Fisher, D.A., Hunt, S.W., III, and Hood, L., 1985, Structure of a gene encoding a murine thymus leukemia antigen, and organization of T1a genes in the Balb/c mouse, *J. Exp. Med.* **162:**528–545.

Flaherty, L., 1976, The T1a region of the mouse: identification of a new serologically defined locus, Qa-2, *Immunogenetics* **3:**533–539.

Gascoigne, N.R.J., Chien, Y., Becker, D.M., Kavaler, J., and Davis, M.M., 1984, Genomic organization and sequence of T cell receptor β constant and joining region genes, *Nature* **310:**387–391.

Gay, D., Coeshott, C., Golde, W., Kappler, J., and Marrack, P., 1986, The major histocompatibility complex-restricted antigen receptor on T cells. IX. Role of accessory molecules in recognition of antigen plus isolated IA, *J. Immunol.* **136:**2026–2032.

Giguere, V., Isobe, K., and Grosveld, F., 1985, Structure of the murine Thy-1 gene, *EMBO J.* **4:**2017–2024.

Gold, D., Clevers, H., Alarcon, B., Dunlaps, S., Novotny, J., Williams, A.F., and Terhorst, C., 1987a, Evolutionary relationship between the CD3 chains of the T lymphocyte receptor complex and the neural glycoprotein N-CAM, *Proc. Natl. Acad. Sci. USA* **84:**7649–7653.

Gold, D.P., van Dongen, J.J.M., Morton, C.C., Bruns, G.A.P., van den Elsen, P., Guerts van Kessel, A.H.M., and Terhorst, C., 1987b, The gene encoding the epsilon subunit of the T3/T-cell receptor complex maps to chromosome 11 in humans and to chromosome 9 in mice, *Proc. Natl. Acad. Sci. USA* **84:**1664–1668.

Gold, D.P., Puck, J.M., Pettey, O.L., Cho, M., Coligan, J., Woody, J.N., and Terhorst, O., 1986, Isolation of cDNA clones encoding the 20K non-glycosylated polypeptide chain of the human T-cell receptor/T3 complex, *Nature* **321:**431–434.

Gottlieb, P.D., 1974, Genetic correlation of a mouse light chain variable region marker with a thymocyte surface antigen, *J. Exp. Med.* **140:**1432–1437.

Greenfield, S., Brostoff, S., Eylar, E.H., and Morell, P., 1973, Protein composition of myelin of the peripheral nervous system, *J. Neurochem.* **20:**1207–1216.

Hayday, A.C., Saito, H., Gillies, S.D., Kranz, D.M.,. Tanigawa, G., Eisen, H.M., and Tonegawa, S., 1985, Structure, organisation and somatic rearrangement of T cell gamma genes, *Cell* **40:**259–269.

He, H.-T., Barbet, J., Chaix, J.-O., and Goridis, C., 1986, Phosphaditylinositol is involved in the membrane attachment of NCAM-120, the smallest component of the neural cell adhesion molecule, *EMBO J.* **5:**2489–2494.

Hedgecock, E.M., Sulston, J.E., and Thompson, J.N., 1983, Mutations affecting programmed cell deaths in the nematode *caenorhabditis elegans, Science* **220:**1277–1279.

Hemperley, J.J., Murray, B.A., Edelman, G.M., and Cunningham, B.A., 1986, Sequence of a cDNA clone encoding the polysialic acid-rich and cytoplasmic domains of the neural cell adhesion molecule N-CAM, *Proc. Natl. Acad. Sci. USA* **83:**3037–3041.

Hill, R.L., Delaney, R., Fellows, R.E., and Lebovitz, H.E., 1966, The evolutionary origins of the immunoglobulins, *Proc. Natl. Acad. Sci. USA* **56**:1762–1769.

Horvitz, H.R., Ellis, H.M., and Sternberg, P.W., 1982, Programmed cell death in nematode development, *Neurosci. Comm.* **1**:56–65.

Howard, F.D., Ledbetter, J.A., Wong, J., Bieber, C.P., Stinson, E.B., and Herzenberg, L.A., 1981, A human T lymphocyte differentiation marker defined by monoclonal antibodies that block E-rosette formation, *J. Immunol.* **126**:2117–2122.

Ishioka, N., Takahashi, N., and Putnam, F.W., 1986, Amino acid sequence of human plasma α_1B-glycoprotein: homology to the immunoglobulin supergene family, *Proc. Natl. Acad. Sci. USA* **83**:2363–2367.

Jefferies, W.A., Green, J.R., and Williams, A.F., 1985, Authentic T helper CD4 (W3/25) antigen on rat peritoneal macrophages, *J. Exp. Med.* **162**:117–127.

Johnson, P., and Williams, A.F., 1986, Striking similarities between antigen receptor J pieces and sequence in the second chain of murine CD8 antigen, *Nature* **323**:74–76.

Johnson, P., Gagnon, J., Barclay, A.N., and Williams, A.F., 1985, Purification, chain separation and sequence of the MRC OX-8 antigen, a marker of rat cytotoxic T lymphocytes, *EMBO J.* **4**:2539–2545.

Joho, R., Weissman, I.L., Early, P., Cole, J., and Hood, L., 1980, Organisation of κ light chain genes in germline and somatic tissue, *Proc. Natl. Acad. Sci. USA* **77**:1106–1110.

Kabat, E.A., Wu, T.T., Bilofsky, H., Reid-Miller, M., and Perry, H., eds., 1983, Sequences of proteins of immunological interest. U.S. Department of Health and Human Services, National Institutes of Health, Bethesda, Maryland.

Kaufman, J.F., Auffray, C., Korman, A.J., Shackelford, D.A., and Strominger, J., 1984, The class II molecules of the human and murine major histocompatibility complex, *Cell* **36**:1–13.

Kranz, D.M., Saito, H., Disteche, C.M., Swisshelm, K., Pravtcheva, D., Ruddle, F.H., Eisen, H.N., and Tonegawa, S., 1985, Chromosomal locations of the murine T cell receptor alpha-chain gene and the T cell gamma gene, *Science* **227**:941–945.

Krissansen, G.W., Owen, M.J., Verbi, W., and Crumpton, M.J., 1986, Primary structure of the T3 γ subunit of the T3/T cell antigen receptor complex deduced from cDNA sequences: evolution of the T3 γ and δ subunits, *EMBO J.* **5**:1799–1808.

Kroczek, R.A., Gunter, K.C., Seligmann, B., and Shevach, E.M., 1986, Induction of T cell activation by monoclonal anti-Thy-1 antibodies, *J. Immunol.* **136**:4379–4384.

Lai, C., Brow, M.A., Nave, K.A., Noronha, A.B., Quarles, R.H., Bloom, F.E., Milner, R.J., and Sutcliffe, J.G., 1987, Two forms of 1B236/myelin-associated ghycoprotein, a cell adhesion molecule for postnatal neural development, are produced by alternative splicing, *Proc. Natl. Acad. Sci. USA* **84**:4337–4341.

Ledbetter, J.A., Seaman, W.E., Tsu, T.T., and Herzenberg, L.A., 1981, Lyt-2 and Lyt-3 antigens are on two different poly-peptide subunits linked by disulfide bonds, *J. Exp. Med.* **153**:1503–1516.

Lemke, G., and Axel, R., 1985, Isolation and sequence of a cDNA encoding the major structural protein of peripheral myelin, *Cell* **40**:501–508.

Lew, A.M., Lillehoj, E.P., Cowan, E.P., Maloy, W.L., Van schravendijk, M.R., and Coligan, J.E., 1986, Class I genes and molecules: an update, *Immunology* **57**:3–18.

Lewis, V.A., Koch, T., Plutner, H., and Mellman, I., 1986, A complementary DNA clone for a macrophage-lymphocyte Fc receptor, *Nature* **324**:372–375.

Littman, D.R.,. Thomas, Y., Maddon, P.J., Chess, L., and Axel, R., 1985, The isolation and sequence of the gene encoding T8: a molecule defining functional classes of T lymphocytes, *Cell* **40**:237–246.

MacDonald, H.R., Glasebrook, A.L., Bron, C., Kelso, A., and Cerottini, J.-C., 1982, Clonal heterogeneity in the functional requirement for Lyt-2/3 molecules on cytolytic T lymphocytes (CTL): possible implications for the affinity of CTL antigen receptors, *Immunol. Rev.* **68**:89–

115.

Maddon, P.J., Littman, D.R., Godfrey, M., Maddon, D.E., Chess, L., and Axel, R., 1985, The isolation and nucleotide sequence of a cDNA encoding the T cell surface protein T4: a new member of the immunoglobulin gene family, *Cell* **42:**93–104.

Mason, D.W., Arthur, R.P., Dallman, M.J., Green, J.R., Spickett, G.P., and Thomas, M.L., 1983, Functions of rat T-lymphocyte subsets isolated by means of monoclonal antibodies, *Immunol. Rev.* **74:**57–82.

McAlpine, P.J., Shows, T.B., Miller, R.L., and Pakstis, A.J., 1985, The 1985 catalog of mapped genes and report of the nomenclature committee cytogenetics and cell genetics, *Human Gene Mapping 8* **40:**8–66.

McCaughan, G.W., Clark, M.J., and Barclay, A.N., 1987a, Characterisation of the human homologue of the rat MRC OX-2 membrane glycoprotein, *Immunogenetics* **25:**329–335.

McCaughan, G.W., Clark, M.J., Hurst, J., Grosveld, F., and Barclay, A.N., 1987b, The gene for MRC OX-2 membrane glycoprotein is localized on human chromosome 3, *Immunogenetics* **25:**133–135.

McLachlan, A.D., 1980, Early evolution of the antibody domain, in: *Protides and Related Subjects 'Protides of the Biological Fluids'*, Volume 28 (H. Peeters, ed.), Pergamon Press, pp. 29–32.

McMichael, A.J., Pilch, J.R., Galfre, G., Mason, D.Y., Fabre, J.W., and Milstein, C., 1979, A human thymocyte antigen defined by a hybrid myeloma monoclonal antibody, *Eur. J. Immunol.* **9:**205–210.

Meuer, S.C., Hussey, R.E., Fabbi, M., Fox, D., Acuto, O., Fitzgerald, K.A., Hodgdon, J.C., Protentis, J.P., Schlossman, S.F., and Reinherz, E.L., 1984, An alternative pathway of T-cell activation: a functional role for the 50 Kd T11 sheep erythrocyte receptor protein, *Cell* **36:**897–906.

Mostov, K.E., Friedlander, M., and Blobel, G., 1984, The receptor for transepithelial transport of IgA and IgM contains multiple immunoglobulin-like domains, *Nature* **308:**37–43.

Murray, B.A., Owens, G.C., Prediger, E.A., Crossin, K.L., Cunningham, B.A., and Edelman, G.M., 1986, Cell surface modulation of the neural cell adhesion molecule resulting from alternative mRNA splicing in a tissue-specific developmental sequence, *J. Cell. Biol.* **103:**1431–1439.

Nakauchi, H., Nolan, G.P., Hsu, C., Huang, H.S., Kavathas, P., and Herzenberg, L.A., 1985, Molecular cloning of Lyt-2, a membrane glycoprotein marking a subset of mouse T lymphocytes: molecular homology to its human counterpart, Leu-2/T8, and to immunoglobulin variable regions, *Proc. Natl. Acad. Sci. USA* **82:**5126–5130.

Nguyen, C., Mattei, M.-G., Mattei, J.-F., Santoni, M.-J., Goridis, C., and Jordan, B.R., 1986, Localization of the human NCAM gene to band q23 of chromosome 11: the third gene coding for a cell interaction molecule mapped to the distal portion of the long arm of chromosome 11, *J. Cell Biol.* **102:**711–715.

Parnes, J.R., Sizer, K.C., Seidman, J.G., Stallings, V., and Hyman, R., 1986, A mutational hotspot within an intron of the mouse β2-microglobulin gene, *EMBO J.* **5:**103–111.

Ravetch, J.V., Luster, A.D., Weinshank, R., Kochan, J., Pavlovec, A., Portnoy, D.A., Hulmes, J., Pan, Y.-C.E., and Unkeless, J.C., 1986, Structural heterogeneity and functional domains of murine immunoglobulin G Fc receptors, *Science* **234:**718–725.

Reif, A.E., and Allen, J.M.V., 1964, the AKR thymic antigen and its distribution in leukemias and nervous tissues, *J. Exp. Med.* **120:**413–433.

Reinherz, E.L., Meuer, S.C., and Schlossman, S.F., 1983, The delineation of antigen receptors on human T lymphocytes, *Immunol. Today* **4:**5–8.

Reinherz, E.L., Kung, P.C., Goldstein, G., and Schlossman, S.F., 1979, Separation of functional subsets of human T cells by a monoclonal antibody, *Proc. Natl. Acad. Sci. USA* **76:**4061–4065.

Richardson, J.S., Richardson, D.C., Thomas, K.A., Silverton, E.W., and Davies, D.R., 1976, Similarity of three-dimensional structure between the immunoglobulin domain and the copper, zinc superoxide dismutase subunit, *J. Mol. Biol.* **102**:221–235.

Rudikoff, S., and Pumphrey, J.G., 1986, Functional antibody lacking a variable-region disulfide bridge, *Proc. Natl. Acad. Sci. USA* **83**:7875–7878.

Rutishauser, U., Hoffman, S., and Edelman, G.M., 1982, Binding properties of a cell adhesion molecule from neural tissue, *Proc. Natl. Acad. Sci. USA* **79**:685–689.

Saito, H., Kranz, D.M., Takagaki, Y., Hayday, A.C., Eisen, H.N., and Tonegawa, S., 1984, Complete primary structure of a heterodimeric T-cell receptor deduced from cDNA sequences, *Nature* **309**:757–762.

Samelson, L.E., Harford, J.B., and Klausner, R.D., 1985, Identification of the components of the murine T cell antigen receptor complex, *Cell* **43**:223–231.

Seki, T., Spurr, N., Obata, F., Goyert, S., Goodfellow, P., and Silver, J., 1985, The human Thy-1 gene: structure and chromosomal location, *Proc. Natl. Acad. Sci. USA* **82**:6657–6661.

Sewell, W.A., Brown, M.H., Dunne, J., Owen, M.J., and Crumpton, M.J., 1986, Molecular cloning of the human T-lymphocyte surface CD2 (T11) antigen, *Proc. Natl. Acad. Sci. USA* **83**:8718–8722.

Shaw, S., Ginther Luce, G.E., Quinones, R., Gress, R.E., Springer, T.A., and Sander, M.E., 1986, Two antigen-independent adhesion pathways used by human cytotoxic T-cell clones, *Nature* **323**:262–264.

Sherr, C.J., Rettenmier, C.W., Sacca, R., Roussel, M.F., Look, A.T., and Stanley, E.R., 1985, The c-*fms* proto-oncogene product is related to the receptor for the mononuclear phagocyte growth factor, CSF-1, *Cell* **41**:665–676.

Solari, R., and Kraehenbuhl, J.F., 1985, The biosynthesis of secretory component and its role in the transepithelial transport of IgA dimer, *Immunol. Today* **6**:17–20.

Steinmetz, M., 1984, Structure, function and evolution of the major histocompatibility complex of the mouse, *TIBS* **9**:224–226.

Sukhatme, V.P., Sizer, K.C., Vollmer, A.C., Hunkapiller, T., and Parnes, J.R., 1985a, The T cell differentiation antigen Leu-2/T8 is homologous to immunoglobulin and T cell receptor variable regions, *Cell* **40**:591–597.

Sukhatme, V.P., Vollmer, A.C., Erikson, J., Isobe, M., Croce, C., and Parnes, J.R., 1985b, Gene for the human T cell differentiation antigen Leu-2/T8 is closely linked to the κ light chain locus on chromosome 2, *J. Exp. Med.* **161**:429–434.

Sutcliffe, J.G., Milner, R.J., Shinnick, T.M., and Bloom, F.E., 1983, Identifying the protein products of brain-specific genes with antibodies to chemically synthesised peptides, *Cell* **33**:671–682.

Swain, S.L., 1983, T cell subsets and the recognition of MHC class, *Immunol. Rev.* **74**:129–142.

Tourvieille, B., Gorman, S.D., Field, E.H., Hunkapiller, T., and Parnes, J.R., 1986, Isolation and sequence of L3T4 complementary DNA clones: expression in T cells and brain, *Science* **234**:610–614.

Trowsdale, J., Young, J.A.T., Kelly, A.P., Austin, P.J., Carson, S., Meunier, H., So, A., Erlich, H.A., Spielman, R.S., Bodmer, J., and Bodmer, W.F., 1985, Structure, sequence and polymorphism in the HLA-D region, *Immunol. Rev.* **85**:5–43.

Tse, A.G.D., Barclay, A.N., Watts, A., and Williams, A.F., 1985, A glycophospholipid tail at the carboxyl-terminus of the Thy-1 glycoprotein of neurons and thymocytes, *Science* **230**:1003–1008.

Unkeless, J.C., 1979, Characterization of a monoclonal antibody directed against mouse macrophage and lymphocyte Fc receptors, *J. Exp. Med.* **150**:580–596.

van den Elsen, P., Shepley, B.-A., Borst, J., Coligan, J.E., Markham, A.F., Orkin, S., and Terhorst, C., 1984, Isolation of cDNA clones encoding the 20K T3 glycoprotein of human T-cell receptor complex, *Nature* **312**:413–418.

van Wauwe, J.P., de Mey, J.R., and Goossens, J.G., 1980, OKT3: a monoclonal anti-human T lymphocyte antibody with potent mitogenic properties, *J. Immunol.* **124**:2708–2713.

Vitetta, E.S., and Capra, J.D., 1978, The protein products of the murine 17th chromosome: Genetics and structure, *Adv. Immunol.* **26**:147–193.

Webb, M., and Barclay, A.N., 1984, Localisation of the MRC OX-2 glycoprotein on the surfaces of neurones, *J. Neurochem.* **43**:1061–1067.

Webb, M., Mason, D.W., and Williams, A.F., 1979, Inhibition of the mixed lymphocyte response with a monoclonal antibody specific for a rat T lymphocyte subset, *Nature* **282**:841–843.

Williams, A.F., 1984, Molecules in the immunoglobulin superfamily, *Immunol. Today* **5**:219–221.

Williams, A.F., 1982, Surface molecules and cell interactions, *J. Theoret. Biol.* **98**:221–234.

Williams, A.F., and Barclay, A.N., 1988, The immunoglobulin superfamily domains for cell surface recognition, *Annu. Rev. Immunol.* **6**:381–405.

Williams, A.F., and Gagnon, J., 1982, Neuronal cell Thy-1 glycoprotein: homology with immunoglobulin, *Science* **216**:696–703.

Williams, A.F., Barclay, A.N., Clark, S.J., Paterson, D.J., and Willis, A.C., 1987, Similarities in sequences and cellular expression between rat CD2 and CD4 antigens, *J. Exp. Med.* **165**:368–380.

Williams, A.F., Barclay, A.N., Clark, M., and Gagnon, J., 1985, Cell surface glycoproteins and the origins of immunity, in: *Proceedings of the Sigrid Juselius Symposium: gene expression during normal and malignant differentiation* (L.C. Anderson, C.G. Gahmberg and P. Ekblom, eds.), Academic Press, pp. 125–138.

Williams, A.F., Galfre, G., and Milstein, C., 1977, Analysis of cell surfaces by xenogeneic myeloma-hybrid antibodies: differentiation antigens of rat lymphocytes, *Cell* **12**:663–673.

Yanagi, Y., Yoshikai, Y., Leggett, K., Clark, S.P., Aleksander, I., and Mak, T.W., 1984, A human T cell-specific cDNA clone encodes a protein having extensive homology to immunoglobulin chains, *Nature* **308**:145–149.

Yarden, Y., Escobedo, J.A., Kuang, W.-J., Yang-Feng, T.L., Daniel, T.O., Tremble, P.M., Chen, E.Y., Ando, M.E., Harkins, R.N., Francke, U., Fried, V.A., Ullrich, A., and Williams, L.T., 1986, Structure of the receptor for platelet-derived growth factor helps define a family of closely related growth factor receptors, *Nature* **323**:226–232.

Yoshikai, Y., Clark, S.P., Taylor, S., Sohn, U., Wilson, B.I., Minden, M.D., and Mak, T.W., 1985, Organisation and sequences of the variable, joining and constant region genes of the human T-cell receptor α-chain, *Nature* **316**:837–840.

4

T-Cell Receptor Genes
Mutant Mice and Genes

DENNIS Y. LOH, MARK A. BEHLKE, AND HUBERT S. CHOU

1. Introduction

To defend against disease, the immune system must be able to recognize a wide variety of foreign antigens with a specificity fine enough to distinguish these foreign antigens from self molecules. For T cells, the T-cell receptor (TcR), a heterodimer composed of an α and a β chain is responsible for the recognition of antigen. Each of the TcR chains is composed of two domains: a constant domain, which is membrane-proximal, and a membrane-distal variable domain, which is responsible for antigen–MHC recognition. The variable domain is encoded by up to three different types of gene segments, which are noncontiguous in the germline; variable (V), joining (J), and at least in the β chain, diversity (D) segments. During T-cell maturation, one of each type is randomly selected and brought together by somatic recombination to form a mature T-cell receptor gene. Thus, the diversity of the T-cell receptor depends on two factors: the number of different gene segments in the germline and the combinatorial diversity generated during the rearrangement process. The murine TcR V_β gene family consists of ~16 V_β subfamilies encoding a total of 20 V_β gene segments (Barth *et al.*, 1985; Behlke *et al.*, 1985). During T-cell ontogeny, these V gene segments undergo somatic DNA rearrangements to D_β–J_β or –J_α gene segments, resulting in a complete V-(D)-J assembly in a fashion similar to the rearrangement process undergone by immunoglobulin genes (Kronenberg *et al.*, 1986).

The organization of the TcR V region genes as multimembered families

DENNIS Y. LOH, MARK A. BEHLKE, and HUBERT S. CHOU ● Departments of Medicine, Microbiology, and Immunology, Howard Hughes Medical Institute, Washington University School of Medicine, St. Louis, Missouri 63110.

confers a number of advantages, such as an increase in sequence diversity and the allowance for functional redundancy among their products. A direct consequence of this redundancy is the enhanced ability of such multigene systems to tolerate deleterious mutations within certain members. The structure of the TcR loci with its linear array of similar genes increases the possibility of germline gene shuffling events such as multiple recombination and gene conversion. In this chapter, we attempt to provide a brief outline of the structure of the murine TcRβ chain gene family and then to discuss the murine strains known to have gross mutations at this locus. We also explore the link between certain mutations and immune system disorders, with reference to the TcRα and γ chains where appropriate.

2. TcR β-Chain Locus

DNA hybridization analysis using V_β-specific probes has shown that the 20 known V_β gene segments fall into 16 distinct subfamilies (Behlke *et al.*, 1985). Moreover, the different V_β subfamilies have been shown to be closely linked on the chromosome (Chou *et al.*, 1987). In fact, 18 out of 20 V_β gene segments have been physically linked using cosmid and phage cloning (Chou and Loh, in press). The amino-acid sequences of the 19 known functional V_β segments are shown in Fig. 1. As is evident from the figure, there are very few blocks of sequence homology among these genes which are suggestive of extended framework areas indicating a wide diversity among these genes. Based on these sequences, we have calculated the homologies between these genes at both DNA and amino-acid levels (Table I).

Overall, members of different subfamilies show relatively low homology (23–64% at the amino-acid level) to each other. Given this high level of heterogeneity, it is intriguing that such a diverse set of polypeptides is able to cover the recognition of myriad antigens in the context of the appropriate major histocompatibility gene products.

3. Deletion-Type Mutants

Two kinds of TcR deletion mutants have been described to date: those involving a large chromosomal deletion and those that are the result of homologous, but unequal, crossovers. Mutants of the former type involve the V_β genes, while an example of the latter type involves the C_β locus.

3.1. V_β Mutants

The first hint of gross genetic differences at the TcR β-chain locus between mouse strains came with the discovery of KJ16 monoclonal antibody by Kappler

Mouse TCR Vβ AA Sequences

```
                  -------- LEADER --------    ----------------------------- V REGION -----------------------------            STD
STD NAME            -20        -10          10        20        30        40        50        60        70        80        90       100
Vβ1      MSCRLLYVSLCLVETALM              NTKITQSPRYLILGRANK-SLECEQHLGHNA-MYWYKQSAEKPPE-LMFLYNLKQLIRNETVP-SRFIPECPDSSKLLLHISAVDPEDSAVYFCASS     B6
Vβ2      MMQFCILCLCVLMASVATDPT           VTLLEQNPRWRLVPRGQAVRLRCICILKNSQYPWMSWYQQDLQKQLQ-WLFTLRSPGDKEVKSLPGADYLATRVTDTELRL--QVAMMSQGRTLYCTCS   B6
Vβ3      MATRLLCYTVLCLLGAGIL             NSKVIQTPRYLVKGQGQKAKMRCIPEKGHPV-VFWYQQNKNNEFKFLINFQNQEVLQQIDMTE-KRFSAECPSNSPCSLEIQSSEAGDSALYLCASS    B6
Vβ4      MGSIFLSCLAVCLLVAGPV             DPKIIQKPKYLVAVTGSEKILICEQYLGHNA-MYWYRQSAKKPLE-FMFSYSYQKLMDNQTAS-SRFQPQSSKKNHLDLQITALKPDDSATYFCASS    B6
Vβ5.1    MSNTAFPDPAWNTTLLSWVALFLLGTSSA   NSGVVQSPRYIIKGKGERSILKCIPISGHLS-VAWYQQTQGQELK-FFIQHYDKMERDKGNLP-SRFSVQQFDDYHSEMNMSALELEDSAVYFCASS    B6
Vβ5.2    MSNTVLADSAWGITLLSWVTVFLLGTSSA   DSGVVQSPRHIIKGKGGRSVLTCIPISGHSN-VVWYQQTLGKELK-FLIQHYEKVERDKGFLP-SRFSVQQFDDYHSEMNMSALELEDSAMYFCASS    B6
Vβ6      MNKWVFCWVTLCLLTVETTHGD          GGIITQTPKFLIGQEGQKLTLKCQQNFNHDT-MYWYRQDSGKGLR-LIYYSITENDLQKGDLS-EGYDASREKKSSSFLTVTSAQKNEMAVFLCASS    B6
Vβ7      VVLCFLGEGLV                     DMKVTQMPRYLIKRMGENVLLECGQDMSHET-MYWYRQDPGLGLQ-LIYISYDVDSNSEGDIP-KGYRVSRKKREHFSLILDSAKTNQTSVYFCASS    B6
Vβ8.1    MGSRLFFVVLILLCAKHM              EAAVTQSPRSKVAVTGGKVTLSCHQTNNHDY-MYWYKQTGHGLR-LIHYSVVADSTEKGDIP-DGYKASRPSQENFSLILELASLSQTAVYFCASS    B6
Vβ8.2    MGSRLFFVLSSLLCSKHM              EAAVTQSPRNKVAVTGGKVTLSCNQTNNHNN-MYWYRQDTGHGLR-LIHYSYGAGSTEKGDIP-DGYKASRPSQENFSLILELATPSQTSVYFCASG    B6
Vβ8.3    MGSRLFLVL-SLLCTKHM              EAAVTQSPRNKVTVTGGNVTLSCRQTNSHNY-MYWYRQDTGHGLR-LIHYSYGAGNLQIGDVP-DGYKATRTTQEDFFLLLELASPSQTSLYFCASS    B6
Vβ9      MDPRLLCCVIFCLLAATFV             DTTVKQNPRYKLARVGKPVNLICSQTMNHDT-MYWYKQDSKKLLK-IMFSYNNKQLIVNETVP-RRFSPQSSDKAHLNLRIKSVEPEDSAVYLCASS    BALB
Vβ10     MGCRLLSCVAFCLLGIGPL             ETAVFQTPNYHYTQVGNEVSFNCKQTLGHDT-MYWYKQDSKKLLK-IMFSYNNKQLIVNETVP-RRFSPQSSDKAHLNLRIKSVEPEDSAVYLCASS    B6
Vβ11     MAPRLLFCLVLCFLRAEPT             NAGVIQTPRHKVTGKGQEATLWCEPISGHSA-VFWYRQTIVQGLEFLTYFRNQAPIDDSG-MPKERFSAQMRNQSHSTLKIQSTQPQDSAVYLCASS   B6
Vβ12     MGIQTLCCVIFYVLIANHT             DAGVTQTPRHEVAEKGQTIILKCEPVSGHND-LFWYRQTKIQGLELLSYFRSKSLMEDGG-AFKDRFKAEMLNSSFSTLKIQPTEPKDSAVYLCASS   B6
Vβ13     GWAVFCLLDTVLS                   EAGVTQSPRYAVLQEGQAVSFWCDPISGHDT-LYWYQQPRDQGPQLLVYFRDEAVIDNSQ-LPSDRFSAVRPKGTNSTLKIQSAKQGDTATYLCASS   B6
Vβ14     MLYSLLAFLLGMFLGVS               AQTIHQMPVAEIKAVGSPLSLGCTIKGKSSPNLYWYWQATGGTLQ-QLFYSITVGQVESV-VQ-LNLSASRPKDQFILSTEKLLLSHSGFYLCAWS    B6
Vβ15     MLLLLLLLGPGCGP                  GALVYQYPRRTICKSGTSMRMECQAVGFQATSVAWYRQSPQKAFELIALSTVNSAIKYEQNFTQEKFPISHPNLSFSSMTVLNAYLEDRGLYLCGA    SJL
Vβ16     MDIWLLGWIIFSFLEAGHI             GPKVLQIPSHQIIDMGQMVTLNCDPVSNHLY-FYWYKQILGQQMEFLVNFYNGKVMEKSKLFK-DQFSVERPDGSYFTLKIQPTALEDSAVYFCASS    SJL
```

Figure 1. Mouse T-cell receptor Vβ amino acid sequences. The amino acid sequences of 19 Vβ genes were translated from the nucleotide sequences of cDNA clones obtained from the inbred mouse strains indicated left of the sequences. A dash (−) indicates a gap in the sequence introduced for alignment. Proposed leader segments are aligned separately.

Table I. Murine V$_\beta$ Homology Matrix[a]

	1	2	3	4	5.1	5.2	6	7	8.1	8.2	8.3	9	10	11	12	13	14	15	16
1	—	26	39	53	40	39	33	39	29	29	32	40	53	37	36	38	24	27	37
2	45	—	35	24	26	23	25	29	28	28	35	24	22	24	25	29	27	22	25
3	53	46	—	33	40	41	35	26	26	24	26	38	35	47	42	39	26	27	36
4	64	46	55	—	36	35	36	34	35	33	32	36	49	38	40	32	29	28	34
5.1	56	48	56	51	—	84	31	36	31	28	29	33	36	43	42	39	32	25	40
5.2	52	47	55	52	89	—	31	34	34	31	32	33	33	43	38	38	28	27	41
6	55	47	51	49	47	46	—	46	48	45	42	35	34	34	33	35	31	29	30
7	56	44	49	53	52	50	56	—	54	53	50	39	33	33	33	37	30	26	32
8.1	50	46	48	49	50	51	57	63	—	89	77	37	37	33	37	38	31	27	37
8.2	50	47	52	48	50	51	56	64	92	—	80	33	33	33	37	35	32	25	32
8.3	49	47	48	50	49	51	56	61	86	88	—	29	35	34	33	34	28	22	28
9	54	47	53	51	56	53	55	53	54	53	55	—	40	31	35	38	27	31	37
10	61	45	54	62	55	55	54	51	51	48	48	57	—	39	34	41	26	28	33
11	57	45	57	56	57	59	48	48	51	52	51	47	55	—	63	52	27	27	43
12	55	47	56	59	52	52	55	50	55	54	54	52	55	72	—	46	25	28	45
13	56	46	54	51	53	53	51	51	52	51	52	54	54	64	61	—	30	23	42
14	48	44	46	46	44	45	49	49	48	47	50	49	45	48	47	48	—	22	27
15	46	47	48	47	48	48	47	43	45	45	45	48	50	48	49	47	46	—	25
16	52	46	55	56	53	52	50	52	48	40	48	49	60	57	58	56	46	46	—

[a]Complete amino acid sequences (above diagonal) and nucleotide sequences (below diagonal) of 19 murine V$_\beta$ genes were aligned and the percent-homology calculated. Leader segments were not included in the comparison.

and Marrack (Haskins *et al.*, 1984). KJ16 was found to stain 15–20% of peripheral T cells in most mouse strains (including BALB/c and C57BL/6), but does not stain T cells from SJL, SWR, C57L, and C57BR mice (Roehm *et al.*, 1985). Thus, it was originally designated as an antiallotypic reagent directed against polymorphic differences in the D_β, J_β, or C_β loci, but further studies began to point towards its recognition of certain V_β gene products. While searching for polymorphism among TcR genes, using all 16 V_β probes, it was found that a full 50% of the V_β gene segments were absent in SJL, SWR, C57L, and C57BR mice, while most other strains contained the full set of known V_β genes (Behlke *et al.*, 1986). Thus, there appeared to be a good correlation between the KJ16⁻ strains and the V_β deletion haplotype. More recently, the analysis of T-cell clones using KJ16 and another anti-TcR antibody, F23.1, suggested that KJ16 recognizes the products of the $V_\beta8.1$ and $V_\beta8.2$ genes, whereas F23.1 recognizes all three members of the $V_\beta8$ subfamily (Behlke *et al.*, 1987).

Although the exact origin of the deletion haplotype is not known, much has been learned since its initial discovery. The four known deletion-mutant strains are identical at the TcR V_β locus by restriction fragment length polymorphism (RFLP) analysis, suggesting that these four seemingly unrelated mouse strains might share a common ancestry. To address this point, we have more recently extended our V_β analysis to the wild mice population and have discovered many independent isolates with the same deletion haplotypes. This suggests that either all four inbred strains originated from the same stock, or from wild type mice that had undergone the deletion, or there is some kind of selection pressure forcing the repeated generation of identical deletion haplotypes in the wild (Huppi and Loh, in press).

As to the nature of the deletion event itself, it was initially proposed that the loss of a contiguous chromosomal segment may account for the deletion haplotype (Behlke *et al.*, 1986). Consistent with this hypothesis, we have physically linked all of the deleted V_β genes in a mouse strain in which the deletion has not occurred (BALB/c) and preliminary data suggests that there are no intervening V_β genes that have not been deleted in this cluster (Chou and Loh, in press).

3.1.1. Immunological/Biological Implications

Several general conclusions can be reached based on present knowledge of the V_β mutant mice:

1. The V_β repertoire is 20 genes in the standard haplotype and ~10 in deletion-haplotype mice. Although the initial proposal of a limited V_α repertoire merely had a statistical basis, no new full-length V_β genes have been discovered in the ensuing 18 months. This is consistent with the limited V_β in the mouse. In addition, analysis of $V_\beta8$-containing cDNA clones shows that they comprise 15–

27% of all C_β^+ cDNAs in the spleen (consistent with the notion that $V_\beta 8$ sub-family products account for KJ16+/F23.1+ T cells). Taken together, these data suggest that V_β repertoire is small in both inbred strains of mice and most wild mice. Furthermore, the same deletion haplotype exists in both wild and inbred mice, whose V_β repertoire is limited to ~10 V_β genes. The small repertoire of the TcR V_β genes limits the ability to generate sequence diversity in the β chains especially in the portion encoded by the V_β gene segment where no somatic hypermutation has been observed (Barth *et al.*, 1985; Behlke *et al.*, 1985; Ikuta *et al.*, 1985).

2. Gross deletion of 50% of V_β genes is apparently well tolerated and implies that there is significant functional redundancy among the V_β gene products against naturally occurring antigens. It is notable that a 50% reduction in the V_β repertoire does not result in overtly immunocompromised animals. Although SJL has well-known immunological abnormalities, including the lack of λ_1 light chains, high IgE levels, and high incidence of reticulum cell sarcoma, there is no obvious causal relationship between the V_β deletion and these abnormalities. Moreover, these abnormalities have not been described in the SWR, C57L, and C57BR deletion-haplotype mice. More interestingly, no known immune unresponsiveness has been mapped in any mouse strain to the β-chain locus on chromosome 6. These data suggest that there is significant functional overlap among the products of the V_β genes that recognize most naturally occurring antigens, the end result of which is that a 50% deletion of the V_β genes does not generate a "hole" in the functional repertoire as measured by standard immunological assays.

It is possible that one may find an immune-responsiveness defect that maps to the β-chain locus, by increasing the specificity of the antigen moiety tested. In fact, it is possible that having the deletion haplotype may be advantageous in nature if the presence of the deleted V_β genes contributes significantly to the development of pathological states (i.e., autoimmunity). This may be one factor that explains why many isolates in the wild mouse population are apparently homozygous for the TcR deletion haplotype.

3.1.2. Possible Applications

3.1.2a. Making anti-V_β antiserum. Because the V_β deletion-haplotype mice lack certain V_β genes, one would expect that products of such V_β genes would be recognized as foreign and would thus be immunogenic. As such, they should be useful in preparing anti-V_β antisera. In fact, Bevan and his colleagues have successfully used such a strategy in developing F23.1, a monoclonal antibody directed against the $V_\beta 8$ family, by immunizing C57L (deletion haplotype) with C57BL/6 (nondeletion haplotype) cells (Staerz *et al.*, 1985).

3.1.2b. Dissection of immune responsiveness to antigens or susceptibility to autoimmunity. Deletion-haplotype mice can be used to study the extent to which immune responsiveness correlates with a given TcR genotype. Susceptibility to autoimmunity may be analysed by appropriate backcrossing of different H-2 and/or TcR haplotype mouse strains.

3.1.2c. Gene transfer experiments. By using deletion-haplotype cell lines or mice as recipients in TcR gene transfer experiments, one is able to avoid background problems stemming from the presence of the endogenous V_β genes. As such, they will be useful in experiments designed to study gene regulation/developmental biological problems.

3.2. C_β-Deletion Mutant

The murine TcR β-chain constant region locus is composed of two tandemly-duplicated units each consisting of one D_β, six J_βs, and one C_β gene segment (Kronenberg *et al.*, 1986). The duplication event must have taken place early in mammalian evolution since virtually identical organization is also seen in man (Toyonaga *et al.*, 1985). This tandem organization allows germline DNA genetic exchange by homologous DNA recombination and/or gene conversion. If one of these gene rearrangements involved a homologous but unequal crossover, it would result in one of the sister chromatids increasing in the gene copy number and its counterpart suffering a loss. While searching for anomalies at the C_β locus, Palmer and his colleagues discovered that the NZW mice had a chromosomal deletion encompassing $C_\beta 1$, $D_\beta 2$, and $J_\beta 2$ gene segments (Kotzin *et al.*, 1985). Soon thereafter, Dixon and his colleagues showed by direct cloning and DNA sequencing that the event leading to the above deletion involved a homologous but unequal crossover between the first exons of $C_\beta 1$ and $C_\beta 2$, resulting in the effective deletion of $C_\beta 1$, $D_\beta 2$, and $J_\beta 2$ gene segments (Noonan *et al.*, 1986). So far, this anomaly of the C_β locus has been limited to the NZW mice, despite extensive screening of other mouse strains, including those that show increased tendency for autoimmunity (MRL, BXSB, and 1pr/1pr strains).

Recently, a new exon ($C_\beta 0$) located $\sim$700 bp upstream of the first exon of $C_\beta 1$ has been described (Behlke and Loh, 1986). This exon is spliced correctly to $C_\beta 1$ in the standard genotype, leading to new mRNA species containing $C_\beta 0$. Furthermore, inclusion of this new exon in the mature mRNA appears to be more common in the thymus than in the spleen. Because the $C_\beta 0$ exon is not deleted in the NZW mouse, the $C_\beta 1$ deletion event effectively juxtaposes $C_\beta 0$ to the first exon of $C_\beta 2$. The result is the correct tissue-specific splicing of $C_\beta 0$ to $C_\beta 2$ in the NZW mice. It is noteworthy that a significantly higher fraction of C_β-containing transcripts contain $C_\beta 0$ exon in the NZW mouse, although its immunological implication is not apparent at the moment.

3.2.1. Immunological/Biological Implications

If the anomaly in the NZW mouse is limited to the deletion of C_β, $D_\beta 2$, and $J_\beta 2$, the primary effect of the deletion event is to reduce the combinatorial diversity to about one-third of that in the wild-type mouse (reduction is not 50% because $D_\beta 1$ can rearrange to both $J_\beta 1$ and $J_\beta 2$ clusters). It is unlikely that this reduction in repertoire is significant unless one postulates that a specific loss of function accompanies the loss of $J_\beta 2$ or $C_\beta 1$ gene segments. Furthermore, NZW V_β genes do not seem to be structurally unusual [two have been sequenced and are identical to their C57BL/6 counterparts, while twelve others show no RFLP (Behlke and Loh, unpublished observation)]. Lastly, as mentioned above, the deletion of $C_\beta 1$ juxtaposes the $C_\beta 0$ exon to $C_\beta 2$ allowing potential expression of a novel mRNA transcript containing $VDJ_\beta C_\beta 0$ spliced correctly to $C_\beta 2$. Although it is tempting to speculate that this new transcript may play an important role in the development of autoimmunity in the F1 (NZW X NZB) mice, no relevant data are available presently.

At present no unifying hypothesis exists to explain the emergence of autoimmunity in the F1 (NZB X NZW), BXSB, and the 1pr/1pr strains. The straightforward hypothesis of linking the pathologic phenotype to the C_β anomaly has not held up since NZW is the only one to show the $C_\beta 1$ deletion. This implies that, if the $C_\beta 1$ deletion is important in the F_1 (NZW and NZB) autoimmune disorder, the same underlying mechanism cannot account for the pathology observed in the other autoimmune mice. The complexity of the genes involved has been recently studied by Yanagi and his colleagues. By backcrossing F_1 (NZB × NZW) to parental strains, they have shown that there are three genetic elements necessary for autoimmunity, including d/z H-2 haplotype, the C_β-deletion haplotype, and the presence of an unknown third gene (Yanagi *et al.*, 1986). Clearly more work is needed to determine the relationship between the autoimmune phenotype and other genetic markers before any reasonable hypothesis can be put forward.

3.3. Vγ-Cγ-Deletion Mutants

The third multigene family whose members undergo gene rearrangements specific to T cells is that of the γ-chain genes. Although the functional role of γ has not been defined, many investigators feel that the γ-han products may be important as part of an alternative T-cell receptor system. The γ-gene system was initially reported to be composed of one or two V_γs with one C_γ, a relatively simple gene system with the capacity for sequence diversity generation only at the functions (Hayday *et al.*, 1985). Recent evidence both in the mouse and human strongly suggests that the γ-chain gene system is more complex with many V_γ and C_γ gene segments.

More recent publications addressing the issue of γ-chain gene organization in the mouse have described genetic heterogeneity of the γ locus in the different strains of mice. For example, in the B10 mouse, the complex consisting of $V_\gamma 3$ (or $V_\gamma 5.7$) and $C_\gamma 3$ is absent (Iwamoto *et al.*, 1986). The biological significance of this heterogeneity is not known.

4. Nondeletion Mutants

Because the TcR genes are organized in multigene families, one expects a proportion of nonfunctional genes as is commonly found among the immunoglobulin gene families. In many cases, the prevalence of such nonfunctional genes may reflect the level of selection put upon the member genes. In this section, we review what we know about such mutant genes to date.

4.1. V_β Genes

In the standard, inbred strains of mice which have ~ 20 V_β gene segments, only one V_β has been found to be nonfunctional. $V_\beta 5.3$, one of the three members of the $V_\beta 5$ family in C56BL/6 mice, is thought to be nonfunctional by a variety of criteria (Chou *et al.*, 1987). All other V_β gene segments appear to be able to encode a mature RNA transcript. In addition, two out of the 14 J_β gene segments appear nonfunctional based on DNA sequence criteria (Kronenberg *et al.*, 1986).

4.2. V_α Genes

At the present time little data are available to realistically estimate the prevalence of pseudogenes among V_α gene segments. This is an important problem since it affects our estimate of the functional V_α repertoire, which is now only based on the number of bands seen on genomic Southern blots. In the one V_α-gene subfamily studied in detail, one of six members of a psuedogene (Chou *et al.*, 1986). Thus, if this proportion of psudogenes is found in V_α families in general, most V_α gene segments may actually contribute to sequence diversity.

4.3. V_γ-Chain

The available data on V_γ reveal that all V_γ genes in mouse are functional, by DNA sequence criteria, whereas four out of nine V_γ gene segments in man are pseudogenes (Garman *et al.*, 1986; LeFranc *et al.*, 1986). It appears that one of the C_γ genes in mouse is a pseudogene because of an insertion (Hayday *et al.*, 1985).

Taken together these numbers suggest that the frequency of pseudogenes is smallest in V_β genes (1/20), intermediate in V_α (1/6), and highest in V_γ (4/9 in human) genes. Whether such numbers reflect the level of selection present or the extent of available redundancy can only be surmised at this point.

References

Arden, B., Klotz, J. L., Siu, G., and Hood, L. E., 1985, Diversity and structure of genes of the α family of mouse T-cell antigen receptor, *Nature* **316**:783.

Barth, R. K., Kim, B. S., Lan, N. C., Hunkapiller, T., Sobieck, N., Winoto, A., Gershenfeld, H., Okada, C., Hansburg, D., Weissman, I. L., and Hood, L., 1985, The murine T-cell receptor uses a limited repertoire of expressed V$_\beta$ gene segments, *Nature* **316**:517.

Behlke, M. A., Spinella, D. G., Chou, H. S., Sha, W., Hartl, D. L., and Loh, D. Y., 1985, T-cell receptor β-chain expression: dependence on relatively few variable region genes, *Science* **229**:566.

Behlke, M. A., Chou, H. S., Huppi, K., and Loh, D. Y., 1986, Murine T-cell receptor mutants with deletions of β-chain variable region genes, *Proc. Natl. Acad. Sci. USA* **83**:767.

Behlke, M. A., and Loh, D. Y., 1986, Alternative splicing of murine T-cell receptor β-chain transcripts, *Nature* **322**:379.

Behlke, M. A., Henkel, T. J., Anderson, S. J., Lan, N. C., Hood, L., Braciale, V. L., Braciale, T. J., and Loh, D. Y., 1987, Expression of a murine polyclonal T-cell receptor marker correlates with the use of specific members of the V$_\beta$8 gene segment subfamily, *J. Exp. Med.* **165**:257.

Chou, H. S., Behlke, M. A., Godambe, S. A., Russell, J. H., Brooks, C. G., and Loh, D. Y., 1986, T-cell receptor genes in an alloreactive CTL clone: implications for rearrangement and germline diversity of variable gene segments, *EMBO J.* **5**:2149.

Chou, H. S., Anderson, S. J., Louie, M. C., Godambe, S. A., Pozzi, M. R., Behlke, M. A., Huppi, K., and Loh, D. Y., 1987, Tandem linkage and unusual RNA splicing of the T-cell receptor β-chain variable-region genes, *Proc. Natl. Acad. Sci. USA*, (in press).

Garman, R. D., Doherty, P., and Raulet, D. H., 1986, Diversity, rearrangement, and expression of murine T-cell gamma genes, *Cell* **45**:733.

Haskins, K., Hannum, C., White, J., Roehm, N., Kubo, R., Kappler, J., and Marrack, P., 1984, The antigen-specific major histocompatibility complex-restricted receptor on T-cells. VI. An antibody to a receptor allotype, *J. Exp. Med.* **160**:452.

Hayday, A., Saito, H., Gillies, S., Kranz, D., Tanigawa, G., Eisen, H. N., and Tonegawa, S., 1985, Structure, organization and somatic rearrangement of T-cell gamma genes, *Cell* **40**:259.

Ikuta, K., Ogura, T., Shimizu, A., and Honjo, T., 1985, Low frequency of somatic mutation in β-chain variable region genes of human T-cell receptors, *Proc. Natl. Acad. Sci. USA* **82**:7701.

Iwamoto, A., Rupp, F., Ohashi, P., Walker, C., Pircher, H., Joho, R., Hengartner, H., and Mak, T. W., 1986, T-cell specific genes in C57BL/10 mice. Sequence and expression of new constant and variable region genes, *J. Exp. Med.* **163**:1203.

Kotzin, B. L., Barr, V. L., and Palmer, E., 1985, A large deletion within the T-cell receptor β-chain gene complex in New Zealand white mice, *Science* **229**:167.

Kronenberg, M., Siu, G., Hood, L., and Shastri, N., 1986, The molecular genetics of the T-cell antigen receptor and T-cell antigen recognition, *Annual Review of Immunology*, Vol. 4 (W. E. Paul, C. G. Fathman, and H. Metzger, eds.), Annual Reviews Inc., Palo Alto, California, pp. 529–591.

LeFranc, M.-P., Forster, A., Baer, R., Stinson, M., and Rabbitts, T., 1986, Diversity and rearrangement of the human T-cell rearranging γ genes: Nine germ-line variable genes belonging to two subgroups, *Cell* **45**:237.

Noonan, D. J., Kofler, R., Singer, P. A., Cardenas, G., Dixon, F. J., and Theofilopoulos, A. N., 1986, Delineation of a defect in T-cell receptor β genes of NZW mice predisposed to autoimmunity, *J. Exp. Med.* **163**:644.

Roehm, N. W., Carbone, A., Kushnir, E., Taylor, B. A., Riblet, R. J., Marrack, P., and Kappler, J. W., 1985, The major histocompatibility complex-restricted antigen receptor on T-cells: the genetics of expression of an allotype, *J. Immunol.* **135**:1276.

Staerz, U. D., Rammensee, H., Benedetto, J. D., and Bevan, M. J., 1985, Characterization of a murine monoclonal antibody specific for an allotypic determinant on T-cell antigen receptor, *J. Immunol.* **134:**3994–4000.

Toyonaga, B., Yoshikai, Y., Vadasz, V., Chin, B., and Mak, T. W., 1985, Organization and sequences of the diversity, joining, and constant region genes of the human T-cell receptor β-chain, *Proc. Natl. Acad. Sci. USA* **82:**8624.

Yanagi, Y., Hirose, S., Nagasawa, R., Shirai, T., Mak, T. W., and Tada, T., 1986, Does the deletion within T-cell receptor β-chain gene of NZW mice contribute to autoimmunity in (NZB × NZW)F$_1$ mice? *Eur. J. Immunol.* **16:**1179.

5

Thymic Ontogeny and the T-Cell Receptor Genes

NICOLETTE CACCIA, ROSANNE SPOLSKI, and TAK W. MAK

The mammalian immune system is composed of a number of well-regulated cells and their products, which provide an effective defense against infection by viruses, bacteria, and parasites. This system can be broadly divided into two interacting components: nonspecific immunity, which is effected by cells such as macrophages and natural killer cells, and specific immunity, which provides the fine tuning and is mediated by B and T lymphocytes. Specific immunity can be further divided into humoral and cell-mediated responses. The humoral response is mediated by B lymphocytes, which secrete immunoglobulins, antigen-specific molecules involved in a number of immune reactions leading to the elimination of antigen (Davies and Metzger, 1983; Honjo, 1983). Regulation of the humoral response and mediation of the cellular response is provided by T lymphocytes, which, unlike B cells, respond exclusively to foreign antigens on the surface of cells, and only recognize these antigens in the context of self class I or class II products encoded by the major histocompatibility complex (MHC). Class I products are found on all cells within an organism, while class II MHC products are expressed only on lymphoid cells and macrophages, and the inability of T cells to respond to antigen except in the context of these products is known as MHC restriction (Zinkernagel and Doherty, 1974; Katz *et al.*, 1973; Rosenthal and Shevach, 1973). T cells participate in a number of cell–cell interactions, which control the differentiation and regulation of the immune response. There are at least two classes of T cells, which perform diverse effector

NICOLETTE CACCIA, ROSANNE SPOLSKI, and TAK W. MAK ● Department of Medicine and Medical Biophysics, Ontario Cancer Institute, University of Toronto, Toronto, Ontario M4X 1K9, Canada.

functions. Helper T cells, for the most part, recognize antigen in the context of class II MHC products and enhance the response of B cells and other T cells. Cytotoxic T cells destroy abnormal host cells, including tumor cells or those infected by virus, and generally recognize antigen in the context of class I MHC molecules.

1. Thymic Ontogeny

The thymus is the major site for the differentiation and maturation of T cells and it is in the thymus that T cells are selected for tolerance to self and 'learn' restriction to self-MHC (Zinkernagel *et al.*, Bevan, 1977; Waldmann *et al.*, 1978). Experiments with chimeric mice have demonstrated that it is the haplotype of the host environment, rather than the genotype of the T cells themselves, that determines MHC restriction (Bevan and Fink, 1978; Fink and Bevan, 1978; Lake *et al.*, 1980; Zinkernagel *et al.*, 1980). This "adaptive differentiation" would seem to be mediated by the thymic environment, which consists of such cells as macrophages and epithelial cells. High levels of MHC class II expression in a number of these cells (Lu, 1980), which have been shown to be closely associated with thymocytes, have led to the suggestion that the association may be instrumental in the maturation of helper T-cell precursors. T-cell maturation may also occur extrathymically, since athymic mice exhibit some T-cell function (Hunig, 1983). The developmental pathway of T cells outside the thymus remains unknown and the significance of extrathymic pathways in normal animals is still being disputed. It has been proposed that within the thymus a selection procedure occurs which only allows the thymocytes with the appropriate specificities to survive (Bevan, 1983). This selection process would account for the high turnover of thymocytes. The majority (95%) of thymocytes die within the thymus, and the actual number which emigrate is quite low (Scollay *et al.*, 1980).

The thymus derives from the epithelio-mesenchymal rudiment, that originates from the third and fourth pharyngeal pouches (LeDourain and Jotereau, 1975). In the fetus T-cell progenitors from the yolk sac and fetal liver migrate to the thymus and begin the first wave of proliferation. In the adult, the thymus continues to receive stem cells from the bone marrow, although the rate at which this occurs is not clear (Enzine *et al.*, 1984). These stem cells undergo considerable proliferation and differentiation under the influence of the thymic microenvironment before export of immunocompetent T cells to the peripheral blood and lymphoid organs.

The thymus is a very heterogeneous organ, which is composed of a variety of cellular subsets. The thymus can be divided morphologically into the cortex and the medulla. The cortex, which accounts for 85–90% of the thymus, is the outer part of the organ. It consists of closely packed small thymocytes, epithelial cells, and macrophages. The medulla, which makes up 10–15% of the thymus, is the central core of the organ and consists of loosely arranged larger thymocytes. The thymus can also be divided on the basis of a number of functional and morphological assays, such as resistance to the drug, hydrocortisone, or the binding of certain subpopulations to sheep red blood cells, or peanut agglutinin (PNA). Since these binding characteristics are dependent on the presence of specific glycoproteins on the cell surface of different thymocytes, these subpopulations can be more precisely defined by the use of monoclonal antibodies against these glycoproteins. Another method of subdividing thymocytes is to evaluate the levels of specific cytoplasmic enzymes that are important in thymocyte development, such as adenosine deaminase (ADA) and terminal deoxynucleotidyl transferase (TdT), whose relative levels vary in different thymocyte subpopulations.

The different subpopulations of thymocytes represent different stages of differentiation; however, the exact lineage relationships between these populations is unclear. Although it is agreed that the discrete stages of thymic ontogeny are accompanied by the acquisition and loss of the various cell-surface and enzymatic markers, the order in which specific markers are gained and lost remains controversial. One of the most significant points of contention is the relationship between the cortex and the medulla in T-cell differentiation. Cells in the medulla appear to be immunocompetent and exhibit cell-surface markers characteristic of peripheral T cells, while the vast majority of cortical cells are phenotypically quite different from peripheral T cells and are functionally immature (Scollay, 1983). It has been shown that the recent emigrants from the thymus are immunocompetent (Scollay et al., 1984a), and are very similar to the cells in the medulla (Scollay et al., 1984b). These facts have led some researchers to postulate a linear differentiation pathway, which proposes that stem cells travel from the bone marow to the cortex, where they proliferate and undergo a strong selection process. They then proceed to the medulla, and finally migrate to the periphery. This theory makes no allowance for pathways that bypass the cortex or medulla, although, it has been demonstrated that in some cases stem cells go directly from the bone marrow to the medulla (Scollay et al., 1980), and, in other cases, that a small subpopulation of cortical cells have been shown to be both immunocompetent and phenotypically mature and to express a receptor molecule involved in lymphocyte homing to peripheral organ (Fink et al., 1985; Reichert et al., 1984). Thus it would seem that there are a number of parallel differentiation pathways within the thymus and that the total picture will encompass a number of different models.

Two current theories of thymic differentiation based on surface markers are shown in Fig. 1. The presence or absence of the surface markers, CD8 and CD4 are used to indicate the maturation of two T-cell subsets. Cytotoxic T-cell activity is usually associated with CD4−CD8+ cells, while the CD4+CD8− phenotype is often characteristic of helper T lymphocytes. The most immature

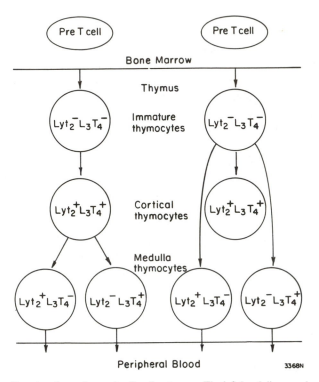

Figure 1. Two alternate schema for murine T-cell ontogeny. The left-hand diagram shows the linear scheme in which a T-cell precursor migrates to the thymus and gives rise to the Lyt2$^-$L3T4$^-$(double negative) population, which then matures into the double positive Lyt2$^+$L3T4$^+$ cortical population that gives rise to Lyt2$^-$L3T4$^+$ and Lyt2$^+$L3T4$^-$ cells, which migrate into the periphery. The second scheme postulates that the Lyt2$^-$L3T4$^+$ populations arise from the double negative cells and that the double positive cells are representative of a maturational deadend L$_3$T$_4$ and Lyt$_2$ have been renamed CD4 and CD8, respectively.

population is that of the double negative CD4$^-$CD8$^-$ thymocytes. One hypothesis of thymic ontogeny postulates that these cells become double positives (CD4$^+$CD8$^+$)before differentiating into either CD4$^+$CD8$^-$ (helper T cells) or CD4$^-$CD8$^+$ (cytotoxic T cells). Another hypothesis proposes that the double negative cells differentiate directly into CD4$^+$CD8$^-$ or CD4$^-$CD8$^+$ cells and the CD4$^+$CD8$^+$ cells represent a maturational 'dead end'. These two hypotheses are shown in Fig. 1.

2. The T-Cell Receptor and Thymic Ontogeny

The immune system of an animal must be able to deal with a wide variety of foreign antigens and to distinguish self from nonself. The required diversity

of the recognition repertoire is mediated by T and B cells, each of which expresses a unique specificity. The processes by which this diversity is generated
have been well studied for immunoglobulin (Davies and Metzger, 1983; Honjo,
1983) and, over the past two years, have been shown to be similar to those used
for the generation of T-cell receptor diversity.

Recognition of antigen by T cells is mediated by the T-cell receptor complex that is comprised of the T-cell receptor, which is responsible for the recognition, and the T3 proteins, which act as signal transducers for the activation of
the T cell (Acuto et al., 1985; Kappler et al., 1983a; Kappler et al., 1983b;
McIntyre and Allison, 1983). The T-cell receptor spans the cell membrane and
is a heterodimer of disulphide-linked α (acidic) and β (basic) chains, which are
composed of variable and constant domains (Chien et al., 1985; Hedrick et al.,
1984a; Edrick et al., 1984b; Saito et al., 1984a; Sim et al., 1984; Yanagi et al.,
1984). The genes encoding these chains are composed of separate, noncontiguous gene segments in the germline, which rearrange to produce a functional
gene, allowing for the generation of a wide variety of receptors by the combinatorial use of these segments. The β-chain variable domain is encoded by variable
(V), diversity (D), and joining (J) segments (Toyonaga et al., 1984), while that
of a α chain has been shown to contain V_α and J_α regions (Winoto et al., 1985;
Yoshikai et al., 1985), and the existence of D_α segments has not been ruled out
although none has been found in the genomic DNA to date.

A third T-cell specific gene that is homologous to immunoglobulin and
rearranges in T cells has been isolated and designated the γ chain (Saito et al.,
1984b). This gene is expressed at high levels in early thymocytes (Raulet et al.,
1985) and at lower levels in helper and cytotoxic T cells, but it has no known
function in T cells as of yet (Heilig et al., 1985; Iwamoto et al., 1986; Raulet et
al., 1985; Saito et al., 1984b; Zauderer et al., 1986). A majority of the γ-chain
transcripts isolated from thymocytes and peripheral blood T cells are nonfunctional, but the encoding sequences are highly conserved (Iwamoto et al., 1986).
The rearrangement and expression of patterns of α-, β-, and γ-chain genes during human ontogeny are shown in Fig. 2. It is evident that much lower levels of
γ-chain expression than those of the α and β chains are seen in both the thymus
and in mature, peripheral blood T cells. In addition, a high level of β-chain
expression is seen in thymocytes when compared to the level in mature T cells
(Yoshikai et al., 1984a).

The rearrangement of T-cell receptor α, β, and γ gene segments is mediated by recognition signals similar to those of immunoglobulin. Immediately
proximal to each of the coding V, D, and J segments is a highly conserved
heptamer, followed by a nonconserved spacer, and then a conserved A/T rich
nonamer. The length of the nonconserved spacer roughly corresponds to either
one (12 nucleotides) or two (23 nucleotides) turns of the DNA helix. It has been
shown that a 12-nucleotide recognition sequence always is involved in recombination with a 23-nucleotide sequence, so that the distribution of these

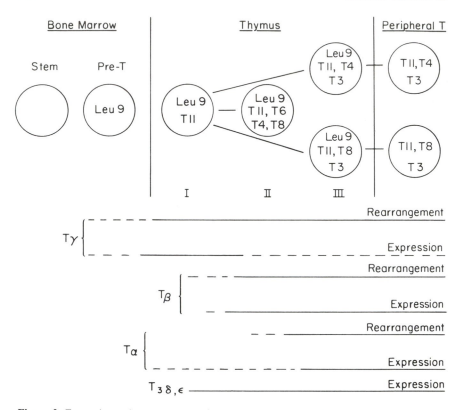

Figure 2. Expression and rearrangement of α, β, and γ chain and T3 genes during human T-cell ontogeny. For rearrangement, the solid lines indicate the stages at which rearrangement can be detected by Southern analysis, while the dashed lines indicate the stages at which rearrangement may occur, but is not detected in the majority of cases. For expression, the solid lines indicate the stages at which expression can be detected by Northern analysis, while the dashed lines indicate the stages during which expression may occur, but only at low levels and/or in a small minority of cells.

two spacers regulates which segments can be joined together and ensures proper joining of V, D, and J segments. The joining process is imprecise, resulting in some variation at which nucleotide the V, D, and J are joined. This variation is augmented by the addition of random nucleotides at the D/J junction by terminal deoxynucleotide transferase. The rearrangement process is closely regulated, with DJ joining preceding the addition of a V segment, and it is not until rearrangement is completed that a functional transcript can be expressed (Yoshikai *et al.*, 1984a).

Since the T-cell receptor is ultimately responsible for the specificity of the individual T cells, the generation of a functional T-cell receptor is crucial in T-cell maturation. An understanding of the mechanisms of the generation process and the role that the receptor plays at different stages of thymic develop-ment will aid in our understanding of the "thymic education" process, including the acquisition of tolerance and the development of MHC restriction. By anal-ogy to B cells, one can expect that different stages of T-cell ontogeny can be related to the generation of a functional T-cell receptor gene from germline segments. Several possible schemes of T-cell receptor gene rearrangement and expression during thymic ontogeny are summarized in Fig. 3. In the first there is prethymic rearrangement and expression of T-cell receptor genes and the thymus acts primarily as an organ of selection to remove cells reactive against self-MHC. While in the thymus other non-TcR changes will occur in these cells such as the acquisition and loss of cell-surface and intracellular differentiation markers. A second model suggests that the T-cell receptor genes are rearranged during thymic ontogeny and selection occurs within the thymus. The third postu-lates an extrathymic pathway for rearrangement and expression of T-cell recep-tor genes. Rearrangement and expression studies suggest that this latter pathway is active in athymic mice (Yoshikai *et al.*, 1986), which exhibit a modicum of learning and selection (Hunig *et al.*, 1983).

In B cells, the rearrangement status of immunoglobulin genes can be used as a marker of B-cell differentiation. It has been found that the immunoglobulin gene rearrangement follows a prescribed sequence, which is linked to matura-tion, with heavy-chain rearrangement preceding that of light-chain and μ-chain rearrangement only occurring when both κ alleles have rearranged nonproduc-tively. Recent studies have shown that rearrangement of α, β, and γ genes in

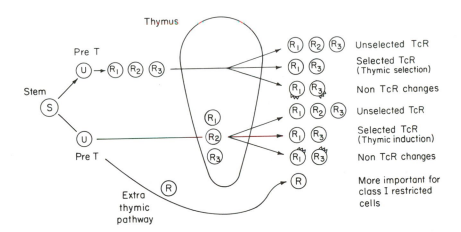

Figure 3. The role of the thymus in T-cell maturation. Three possible pathways of T-cell selection based on the expression of T-cell receptor specificities; prethymic selection, intrathymic selection, and extrathymic selection. U, unrearranged genes; R, rearranged genes. The subscripted numbers denote different rearrangements.

maturing T cells exhibits a similar sequence, which is linked to the different developmental stages.

If one looks at the expression of T-cell receptor messages in the murine thymus during embryogenesis, γ-chain mRNA appears first followed by β-chain, and then α-chain messages (Haars *et al.*, 1986; Raulet *et al.*, 1985; Snodgrass *et al.*, 1985a: Snodgrass *et al.*, 1985b). The earliest stage of thymic development studied was the day-14 fetal thymus, which contains developing T cells that are mainly Thy-1[+] and CD4[-]CD8[-] (Ceredig *et al.*, 1983b; Habu and Okamura, 1984; van Ewijk *et al.*, 1982) with low levels of IL-2 receptor expression (Habu *et al.*, 1985; Takacs *et al.*, 1984). These cells express γ (Haars *et al.*, 1986) but not β or α transcripts. By day 15, the short 1.0-kb β message is in evidence with only low levels of the longer 1.3-kb transcript (Haars *et al.*, 1986; Snodgrass *et al.*, 1985b). By day 16, α-chain transcription begins to be seen and γ levels have decreased significantly. Elevated levels of β-chain expression, compared to those in mature T cells, are first seen on day 16 and continue until birth (Raulet *et al.*, 1985; Haars *et al.* 1986; Snodgrass *et al.*, 1985b; Widmer *et al.*, 1981), whereas expression of the α chain only reaches significant levels on day 17 (Raulet *et al.*, 1985) when surface T-cell receptor is first seen (Ceredig *et al.*, 1983; Haars *et al.*, 1986; Mathieson *et al.*, 1981; Roehm *et al.*, 1984). The observation that the γ chain shows the opposite expression pattern to that of the α chain, with high levels at day 14, which decline by the 17th day (Born *et al.*, 1985; Raulet *et al.*, 1985), has led to the suggestion that the γ chain is used in place of the α chain in the fetus, in a scheme similar to that of the fetal and adult globin genes (Raulet *et al.*, 1985) but the recent discovery of a second T-cell receptor composed of γ and a new chain, δ, raises the possibility that $\gamma\delta$[+] cells are important early in T-cell ontogeny and that only with the rearrangement of α does β get expressed on the cell surface.

The pattern of β-chain rearrangement in the thymus is consistent with the expression data. Born *et al.* (1985) produced thymocyte hybridomas from cells derived from various stages of murine ontogeny. Examination of these hybidomas showed a steady increase in the frequency of rearranged β complexes during thymic development from days 14–17 with diversity segment to joining segment ($D_\beta J_\beta$) rearrangement preceding the other types. Rearrangement involving the first cluster of genes, $D_{\beta 1}J_{\beta 1}$, were seen earlier than those involving the second cluster, $D_{\beta 2}J_{\beta 2}$, until their levels were equal at day 17. The complexity of the rearrangements also increased during development. There was only evidence of substantial rearrangement involving variable segments, to produce complete rearrangements on day 16, one day prior to significant T-cell receptor expression on the cell surface (Snodgrass *et al.*, 1985b; Widmer *et al.*, 1981). These results imply that the development of thymocytes during ontogeny is coupled with rearrangement of T-cell receptor genes, indicating perhaps that the thymus is not merely an organ of selection, but also a site for the development of recognition at a molecular level. More recent data from this same group of workers (Born *et al.*, 1986) suggests that γ- and β-chain genes are expressed more or less synchronously.

This same linkage betwwen thymocyte development and changes in T-cell receptor genes have also been demonstrated in the postnatal human thymus. The β chain is expressed at a much higher level in thymocytes than in peripheral blood T cells (Yoshikai *et al.*, 1984b), with this increased level sustained through intrathymic differentiation, while the α chain is only expressed in the more mature T3+ thymocytes, but at the same level as in peripheral blood cells (Collins *et al.*, 1985).

3. Thymic Leukemias

Our understanding of human thymic development has been aided by the study of thymic leukemias. The classification of these leukemias is based upon their surface markers and levels of specific intracellular enzymes. Human T-cell leukemias have been extensively analyzed (Minowada *et al.*, 1983), and a number of different leukemia subsets, representative of the various thymus subpopulations, have been isolated. These leukemias, which would seem to be derived from thymocytes arrested at a given stage of development as a result of the malignant transformation, provide a convenient source of a homogenous cell population, from which stable permanent cell lines can be established. The greatest advantage of studying leukemic cell lines is that they provide an expanded monoclonal population, which is easier to analyze than a heterogeneous one. Their cell-surface marker profiles are relatively stable in long term culture and are characteristic of different thymocyte subpopulations. The markers appear to be normal gene products which are not tumor-specific. Another advantage of using these lines is that they can often be induced to exhibit a more mature phenotype by phorbol esters (Nagasawa and Mak, 1980) and so can be used to study differentiation.

A number of thymic differentiation schema have been proposed by researchers, based on their work with leukemic T-cell lines, that classify these T-cell lines in a relative order of differentiation based on their surface phenotype and enzyme markers. These schemes divide the T-cell lineage into two to five stages, to which most of the T-cell lines can be assigned. One such scheme proposed by Minowada, is based on the presence or absence of a number of cell-surface antigens and enzymes, such as the sheep red blood cell receptor, cALL, the TAg and terminal deoxynucleotidyl transferase (TdT) (Minowada *et al.*, 1981, 1982, 1983).

The differentiation scheme proposed by Reinhertz *et al.* (1980) assumes that as T cells differentiate they pass from the cortex to the medulla, and then to the periphery, and is based on the presence of cell surface glycoproteins defined by monoclonal antibodies, such as T1, T3, T4, T5, T6, T8, T10, and T11, (CD2). T1, T3 (CD3), and T10 antigens are used as maturation markers, as T1 and T3 are present on all peripheral T cells, and are supposed to be present on all mature thymocytes, while the T10 antigen is found on thymocytes, but not on peripheral T cells. T4, T5, and T8 are used to define precursors of functional subsets, since in peripheral T cells, the T4 (CD4) antigen is found on the helper/ inducer subset, while the T5 and T8 (CD8) antigens are found on the suppressor/ cytotoxic T cells. The cortical antigen, T6, is used to define cells from that region.

Royer *et al.* (1984) have divided thymocytes into three subgroups on the basis of Reinhertz' classification system. Stage I thymocytes (T11+,T6−, T3−) are cortical and comprise about 10% of the total population; stage II thymocytes (T11+,T6+,T3−) are also cortical, but they comprise the bulk of thymocytes (60–70%); stage II thymocytes (T+,T6−,T3+) are functionally competent and are comprised of two subpopulations, expressing either T4 or T8 in addition to T3 and T11. Rearrangement of β-chain genes is evident in stage II and III thymocytes, whereas surface expression of a α and β chains is only seen in the stage III thymocytes (Royer *et al.*, 1984, 1985). In Stage I, levels of β-chain transcripts are minimal, and α-chain transcription is virtually undetectable, while in stage II, β-chain transcription is maximal and levels of α-chain message low, and in stage III α- and β-chain messages are at a level comparable to that of peripheral blood T cells (Royer *et al.*, 1985). The high levels of β messages, which precede significant levels of α-chain transcripts, suggest that α-chain expression or stability may be regulated by β-chain expression. This hypothesis is supported by the finding that in a mutant cell line that was not capable of making a functional β-chain message, levels of α-chain message were greatly diminished when compared to those in the parent line (Ohashi *et al.*, 1985).

An analysis of 14 human thymic leukemia cell lines and 15 functional T-cell lines support this hypothesis (Sangster *et al.*, 1986). Both β- and γ chain rearrangement were seen in all the functional T-cell lines and in all but one of the thymic leukemia cell lines.The unrearranged line may represent a very immature stage in thymic development or even a cell that has not yet committed to the T-cell lineage. Expression of β-chain message was found in all lines that exhibited rearrangement, while the α-chain was expressed in all the functional T-cell lines, but only in $\frac{9}{10}$ of the thymic leukemia cell lines. Based on these results, there would seem to be an ordered rearrangement and expression of α-, β-, and γ-chain genes during T-cell development with β- and γ-chain rearrangement and expression preceding those of the chain.

4. Conclusions

Despite the fact that the thymus plays an important role in the selection and education of T cells, until recently this organ remained very much a black box. The use of T-cell receptor probes in the study of thymic ontogeny has provided a candle to investigate the processes within this "box". These results suggest that T-cell ontogeny, like that of B cells, is characterized by an orderly series of steps where γ-chain genes are transcribed and rearranged, followed closely, if not simultaneously, by $D_\beta J_\beta$ rearrangement and expression, which progresses to rearrangement and transcription of V_β sequences, and finally the α-chain genes are rearranged and transcribed.

Although a number of questions still remain, including whether cortical and medullary cells are part of the same differentiation pathway or whether they follow different maturation routes, and the role sequential thymic expression of T-cell receptor genes plays in T-cell selection, these results provide a framework for investigation of these questions. One of the most interesting observations is that athymic mice express a high level of functional γ-chain message, with drastically reduced levels of α- and β-chain transcripts (Yoshikai et al., 1986). This suggests that there may be a separate lineage of T cells that do not require a thymic environment in which to mature. This exciting possibility is illustrated schematically in Fig. 4. Thus, it is possible that, while α- and β-genes are rearranged and expressed during thymic development, the γ-chain genes can be rearranged and expressed in a prethymic or extrathymic pathway (see Fig. 3). Thus, it is possible that the receptor containing the γ chain represents a second recognition structure that is active in a separate T-cell pathway, in which the α- and β-chain TcR may not be used. Although no functional α- and β-chain transcripts are detected in young (8 weeks old) athymic mice, a low level of func-

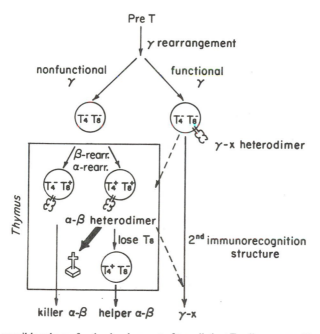

Figure 4. A possible scheme for the development of two distinct T-cell receptors. Two pathways of T-cell development are illustrated. Conventional thymic development is shown on the left, with α- and β-chain rearrangement and expression occurring in the thymus, after the nonfunctional rearrangement of the γ chain prethymically. The extrathymic pathway is shown on the right. This pathway would be followed by cells with functional γ-chain rearrangement and could involve a sojourn in the thymus.

tional α- and β-chain messages can be detected in these mice later in their development (20 weeks) (Kishihara *et al.*, 1987). These findings are consistent with the hypothesis that rearrangement of α, β, and γ genes can occur in the absence of a proper thymic environment.

References

Acuto, O., Hussey, R.E., Fitzgerald, K. A., Protentis, J. P., Meuer, S. C., Schlossman, S. F., and Reinherz, E. L., 1985, The human T cell receptor: appearance in ontogeny and biochemical relationship of α and β subunits on IL-2 dependent clones and T cell tumors, *Cell* **34**:717–726.

Bevan, M. J., 1977, In a radiation chimaera, host H-2 antigens determine immune responsiveness of donor cytotoxic cells, *Nature* **269**:417–418.

Bevan, M. J., 1983, Thymic education. *in T Lymphocytes Today* (J. R. Inglis, ed.), pp. 38–41, Elsevier Science Publishers, Amsterdam.

Bevan, M. J., and Fink, P. J., 1978, The influence of thymus H-2 antigen on the specificity of maturing killer and helper cells, *Imm. Rev.* **42**:3–19.

Born, W., Rathburn, G., Tucker, P., Marrack, P., and Kappler, J., 1986, Synchronized rearrangement of T cell γ and β chain genes during fetal thymocyte development, *Science* **234**:479–482.

Born, W., Yague, J., Palmet, E., Kappler, J., and Marrack, P., 1985, Rearrangement of T-cell receptor β-chain genes during T cell development, *Proc. Natl. Acad. Sci. U.S.A.* **82**:2925–2929.

Ceredig, R., Dialynis, D. P., Fitch, F. W., and MacDonald, H. R., 1983a, Precursors of T cell growth factor producing cells in the thymus, *JEM* **158**:1654–1671.

Ceredig, R., MacDonald, H. R., and Jenkinson, E. J., 1983b, Flow microfluorometric analysis of mouse thymus development in vivo and in vitro, *Eur. J. Immunol.* **13**:185–190.

Chien, Y., Becker, D., Lindsten, T., Okamura, M., Cohen, D., and Davis, M., 1985, A third type of murine T-cell receptor gene, *Nature* **312**:31–36.

Collins, M. K. L., Tanigawa, G., Kissonherghis, A.-M., Ritter, M., Price, K. M., Tonegawa, S., and Owen, M.J., 1985, Regulation of T-cell receptor gene expression in human T-cell development, *Proc. Natl. Acad. Sci. USA* **82**:4503–4507.

Davies D. R., and Metzger, H., 1983, Structural basis of antibody function, *Annu. Rev. Immunol.* **1**:63–86.

Enzine, S., Weissman, I. L., and Rouse, R. V., 1984, Bone marrow cells give rise to distinct cell clones within the thymus, *Nature* **309**:629–631.

Fink, P. J., and Bevan, M. J., 1978, H-2 antigens of the thymus determine lymphocyte specificity, *J. Exp. Med.* **148**:766–775.

Fink, P. J., Gallatin, W. M., Reichert, R. A., Butcher, E. C., and Weissman, I. L., 1985, Homing receptor-bearing thymocytes, an immunocompetent cortical subpopulation, *Nature* **313**: 233–235.

Haars, R., Kronenberg, M., Owen, F., Gallatin, M., Weissman, I., and Hood, L., 1986, Rearrangement and expression of T-cell antigen receptor and γ chain genes during thymic differentiation, *J. Exp. Med.* **164**:1–24.

Habu, S., and Okamura, K., 1984, Cell surface antigen marking the stages of murine T cell ontogeny and its functional subsets, *Immunol. Rev.* **82**:117–139.

Habu, S., Okamura, K., Diamentstein, T., and Shevach, E. M., 1985, Expression of interleukin 2 receptor on murine fetal thymocytes, *Eur. J. Immunol.* **15**:456–460.

Hedrick, S. M., Cohen, D. I., Nielsen, E. A., and Davis M. M., 1984a, Isolation of cDNA clones encoding T cell specific membrane-associated proteins, *Nature* **308**:149–152.

Hedrick, S. M., Neilsen, E. A., Kavaler, J., Cohen, D. I., and Davis, M. M., 1984b, Sequence relationships between putative T-cell receptor polypeptides and immunoglobulins, *Nature* **308**:153–158.

Heilig, J. S., Glimcher, L. H., Kranz, D. M., Clayton, L. K., Greenstein, J. L., Saito, H., Maxam, A. M., Burakoff, S. J., Eisen, H. N., and Tonegawa, S., 1985, Expression of the T-cell specific γ gene is unnecessary in T cells recognizing class II MHC determinants, *Nature* **317**:68–70.

Honjo, T., 1983, Immunoglobulin genes, *Annu. Rev. Immunol.* **1**:499–528.

Hunig, T., 1983, T cell function and specificity in athymic mice, in: *T Lymphocytes Today* (J. R. Inglis, ed.), pp. 42–45, Elsevier Science Publisher, Amsterdam.

Iwamoto, I., Ohashi, P., Walker, C., Rupp, F., Yoho, H., Hengartner, H., and Mak, T. W., 1986, The murine γ chain genes in B10 mice: Sequence and expression of new constant and variable genes, *J. Exp. Med.* **163**:1203–1212.

Kappler, J., Kubo, R., Haskins, K., Hannum, C., Marrack, P., Pigeon, M., McIntyre, B., Allison, J., and Trowbridge, I., 1983a, the MHC-restricted antigen receptor on T cells: identification of constant and variable peptides, *Cell* **35**:295–302.

Kappler, J., Kubo, R., Haskins, K., White, J., and Marrack, P., 1983b, The mouse T cell receptor: comparison of MHC-restricted receptors on two T cell hybridomas, *Cell* **34**:727–737.

Katz, D. H., Hamaoka, T., and Benacerraf, B., 1973, Cell interactions between histoincompatible T and B lymphocytes, *J. Exp. Med.* **137**:1405–1418.

Kishihara, K., Yoshikai, Y., Matsuzakai, G., Mak, T. W., Nomoto, K., 1987, Functional α and β chain T cell receptor messages can be detected in old, but not young, athymic mice, *Eur. J., Immunol.* (in press).

Lake, J. P., Andrew, M. E., Pierce, C. W., and Braciale, T. J., 1980, Sendai virus-specific, H-2 restricted cytotoxic T lymphocyte responses of nude mice grafted with allogeneic or semi-allogeneic thymus glands, *J. Exp. Med.* **152**:1805–1810.

LeDourain, N. M., and Jotereau, F. V., 1975, Tracing of cells in the avian thymus through embryonic life in interspecific chimeras, *J. Exp. Med.* **142**:17–40.

Lu, C. Y., Beller, D. I., and Unanue, E. R., 1980, During ontogeny, Ia bearing accessory cells are found early in the thymus, but late in the spleen, *Proc. Natl. Acad. Sci. USA* **77**:1597–1601.

McIntyre, B. W., and Allison, J. P., 1983, The mouse T cell receptor: structural heterogeneity of molecules of normal T cells defined by xenoantiserum, *Cell* **34**:739–746.

Mathieson, B. J., Sharrow, S. O., Rosenberg, Y., and Hammerling, U., 1981, Lyt1+23- cells appear in the thymus before Lyt 123+ cells, *Nature* **289**:179–181.

Minowada, J., Koshiba, H., Sagawa, K., Kubonishi, I., Lok, M. S., Tatsumi, E., Han, T., Srivastava, B. I. S., and Ohnuma, T., 1981, Marker profiles of human leukemia and lymphoma cell lines, *J. Cancer Res. Clin. Oncol.* **101**:91–100.

Minowada, J., Minato, K., Srivastava, B. I. S., Nakazawa, S., Kubonishi, I., Tatsumi, E., Ohnuma, T., Ozer, H., Freeman, A. I., Henderson, E. S., and Gallo, R. C., 1982. A model scheme of human hematopoietic cell differentiation as determined by leukemia-lymphoma study: T-cell lineages, in: *Current Concepts in Human Immunology and Cancer Immunomodulation* (Serrou, B., Rosenfeld, C., Daniels, J. C., and Saunders, J. P., eds.), Elsevier Biomedical Press B. V.

Minowada, J., Minato, K., Tatsumi, E., Sugimoto, T., Nakazawa, S., Ohnuma, T., Kubonishi, I., Miyoshi, I., Frankel, A., and Gallo, R. C., 1983. A model scheme for human hematopoietic cell differentiation as determined by multiple markers of leukemia-lymphomas, In: *Haematology and Blood Transfusion*, Vol. 28 (Neth, R., Gallo, R., Greaves, M., Moore, O., and Winkler, F., eds.), pp. 348–350, Springer-Verlag, Berlin.

Nagasawa, K., and Mak, T.W., 1980, Phorbol esters induce differentiation in human malignant T lymphoblasts, *Proc. Natl. Acad. Sci. USA* **77**:2964–2968.

Ohashi, P., Mak, T. W., Van den Elsen, P., Yanagi, Y., Yoshikai, Y., Calman, A. F., Terhorst, C., Stobo, J. D., and Weiss, A., 1985, Reconstitution of an active T3/T cell antigen receptor in human T cells by DNA transfer, *Nature* **316**:602–606.

Raulet, D. H., Garman, R. D., Saito, H., and Tonegawa, S., 1985, Developmental regulation of T-cell receptor gene expression, *Nature* **314:**103–107.

Reichert, R. A., Gallatin, W. M., Butcher, E. C., and Weissman, I. L., 1984, A homing receptor-bearing cortical thymocyte subset: implications for cortisone-resistant thymocytes, *Cell* **38:**89–99.

Reinhertz, E. L., Kung, P. C., Goldstein, G., Levey, R. H., and Schlossman, S. F., 1980, Discrete stages of human intrathymic differentiation: Analysis of normal thymocytes and leukemia lymphoblasts of T-cell lineage, *Proc. Natl. Acad. Sci. U.S.A.* **77:**1588–1592.

Roehm, N., Herron, L., Cambier, J., DiGuisto, D., Haskins, K., Kappler, J., and Marrack, P., 1984, the major histocompatability complex-restricted antigen receptor on T cells: Distribution in thymus and peripheral blood T cells, *Cell* **38:**577–584.

Rosenthal, A. S., and Shevach, E. M., 1973, Function of macrophages in antigen recognition by guinea pig T lymphocytes, *J. Exp. Med.* **138:**1194–1212.

Royer, H. D., Acuto, A., Fabbi, M., Tizzard, R., Rammachandran, K., Smart, J., and Reinherz, E. L., 1984, Genes encoding the Ti β subunit of the antigen/MHC receptor undergo rearrangement during intrathymic ontogeny prior to surface T3-Ti expression, *Cell* **39:**261–266.

Royer, H. D., Ramarli, D., Acuto, O., Campen, T. J., and Reinherz, E. L., 1985, Genes encoding the T-cell receptor α and β subunits are transcribed in an ordered manner during intrathymic ontogeny, *Proc. Natl. Acad. Sci. USA* **82:**5510–5514.

Saito, H., Kranz, D., Takagaki, Y., Hayday, A., Eisen, H., and Tonegawa, S., 1984a, A third rearranged and expressed gene in a clone of cytotoxic T lymphocytes, *Nature* **312:**36–40.

Saito, H., Kranz, D. M., Takagaki, Y., Hayday, A., Eisen, A. C., Tonegawa, S., 1984b, Complete primary structure of a heterodimeric T-cell receptor deduced from cDNA sequences, *Nature* **309:**757–762.

Sangster, B., Minowada, J., Suci-Foca, N., Minden, M., and Mak, T. W., 1986, Rearrangement and expression of the α, β and γ T cell receptor genes in human leukemias and functional T cells, *J. Exp. Med.* **163:**1491–1507.

Scollay, R., 1983, Intrathymic events in the differentiation of T lymphocytes: a continuing enigma, in: *T Lymphocytes Today* (J. R. Inglis, ed.), pp. 52–56, Elsevier Science Publishers, Amsterdam.

Scollay, R. G., Butcher, E. C., and Weissman, I. L., 1980, Quantitative aspects of cellular traffic from the thymus to the periphery in mice, *Eur. J. Immunol.* **10:**210–218.

Scollay, R., Chen, W.-F., and Shortman, K., 1984a, The functional capabilities of cells leaving the thymus, *J. Immunol.* **132:**25–30.

Scollay, R., Wilson, A., and Shortman, K., 1984b, Thymus cell migration: analysis of thymus emigrants with markers that distinguish medullary thymocytes from peripheral T cells, *J. Immunol.* **132:**1089–1094.

Sim, G. K., Yague, J., Nelson, J., Marrack,P., Palmer, E., Augustin, A., and Kappler, J., 1984, Primary structure of human T-cell receptor α chain, *Nature* **312:**771–775.

Snodgrass, H. R., Dembic, Z., Steinmetz, M., von Boehmer, H., 1985a, Expression of T-cell antigen receptor genes during fetal development in the thymus, *Nature* **313:**232–233.

Snodgrass, H. R., Kisielow, P., Kiefer, M., Steinmetz, M., and von Boehmer, H., 1985b, Ontogeny of the T-cell antigen receptor within the thymus, *Nature* **313:**592–595.

Takacs, L., Osawa, H., and Diamentstein, T., 1984, Detection and localization by the monoclonal anti-interleukin-2 receptor antibody AMT-13 of IL2 receptor-bearing cells in the developing thymus of the mouse embryo and in the thymus of cortisone-treated mice, *Eur. J. Immunol.* **14:**1152–1156.

Toyonaga, B., Yanagi, Y., Suciu-Foca, N., Minden, M., and Mak, T. W., 1984, Rearrangements of T cell receptor gene YT35 in human DNA from thymic leukemia T cell lines and functional T cell clones, *Nature* **311:**385–387.

van Ewijk, W., Jenkinsson, E. J., Owen, J. J. T., 1982, Detection of Thy-1, T-200 Lyt-1 and Lyt-2 bearing cells in the developing lymphoid organs of the mouse embryo in vivo and in vitro, *Eur. J. Immunol.* **12:**262–271.

Waldmann, H., Pope, H., Brent, L., and Bighouse, K., 1978, Influence of the major histocompatibility complex on lymphocyte interactions in antibody formation, *Nature* **274:**166–168.

Widmer, M. D., MacDonald, H. R., and Cerottini, J. C., 1981, Limiting dilution analysis of alloantigen-reactive T lymphocytes, VI, Ontogeny of cytolytic T lymphocyte precursors in the thymus, *Thymus* **2:**245–255.

Winoto, A., Mjolsness, S., and Hood, L., 1985, Genomic organization of the genes encoding mouse T-cell receptor α chain, *Nature* **316:**832–836.

Yanagi, Y., Yoshikai, Y., Leggett, K., Clark, S. P., Aleksander, I., Mak, T. W., 1984, A human T cell specific cDNA clone encodes a protein having extensive homology to immunoglobulin chains, *Nature* **308:**145–149.

Yoshikai, Y., Anatoniou, D., Clark, S. P., Yanagi, Y., Sangster, R., Van den Elsen, P., Terhorst, C., and Mak, T. W., 1984a, Sequence and expression of transcripts of the human T-cell receptor β-chain genes, *Nature* **312:**521–524.

Yoshikai, Y., Clark, S. P., Taylor, S., Sohn, U., Wilson, B., Minden, M., and Mak, T. W., 1985, Organization and sequences of the variable, joining and constant region genes of the human T cell receptor α chain, *Nature* **31:**837–840.

Yoshikai, Y., Reis, M. D., and Mak, T. W., 1986, Athymic mice express a high level of functional γ chain, but drastically reduced levels of α and β chain T cell receptor messages, *Nature* **324:**482–485.

Yoshikai, Y., Yanagi, Y., Suciu-Foca, N., and Mak, T. W., 1984b, Presence of T cell receptor mRNA in functionally distinct T cells and elevation during intrathymic differentiation, *Nature* **310:**506–508.

Zauderer, M., Iwamoto, I., and Mak, T. W., 1986, Gamma gene expression in autoreactive helper T cells, *J. Exp. Med.* **163:**1314–1318.

Zinkernagel, R. M., Althage, A., Waterfield, E., Kindred, B., Welsh, R. M., Callahan, G., and Pincetl, P., 1980, Restriction specificities, alloreactivity, and allotolerance expressed by T cells from nude mice reconstituted with H-2 compatible or incompatible thymus grafts, *J. Exp. Med.* **151:**376–399.

Zinkernagel, R. M., and Doherty, P. C., 1974, Restriction of in vitro T cell-mediated cytotoxicity in lymphocytic choriomeningitis within a syngeneic or semiallogeneic system, *Nature* **248:**701–702.

Zinkernagel, R. M., Gallahan, G. N., Althage, A., Cooper, S., Klein, P. A., and Klein, J., 1978, On the thymus in the differentiation of H-2 self recognition by T cells: Evidence for dual recognition? *J. Exp. Med.* **147:**882–896.

6

The T-Cell Receptor/T3 Complex on the Surface of Human and Murine T Lymphocytes

COX TERHORST, BENJAMIN BERKHOUT,
BALBINO ALARCON, HANS CLEVERS,
KATIA GEORGOPOULOS, DANIEL GOLD,
HANS OETTGEN, CAROLYN PETTEY,
PETER VAN DEN ELSEN, and TOM WILEMAN

1. T-Cell Receptors for Antigen

Thymus-derived lymphocytes (T cells) are essential in the protection of verte-brates from environmental pathogens. As effector cells, they serve to recognize and eliminate cells bearing infectious agents and tumor antigens. In addition, they play a major role in the functional regulation of the immune response. Unlike B lymphocytes, which mediate their antigen-specific functions via se-creted or membrane bound immunoglobulins that can bind to soluble ligands, T cells are responsive only to cell-surface antigens. Specifically, T-lymphocyte membrane receptors recognize antigen presenting cells (APCs), which express the autologous class I or class II integral membrane protein products of the major histocompatibility complex (MHC). This interaction between cells can lead to a variety of responses by the antigen-specific T lymphocytes, including proliferation, production of lymphokines, and target cell cytolysis.

The corecognition of antigen and autologous MHC-encoded proteins (asso-ciative recognition) by T cells as well as responsiveness to allogeneic MHC

COX TERHORST, BENJAMIN BERKHOUT, BALBINO ALARCON, HANS CLEVERS, KATIA GEORGOPOULOS, DANIEL GOLD, HANS OETTGEN, CAROLYN PETTEY, PETER VAN DEN ELSEN, and TOM WILEMAN • Laboratory of Molecular Immunology, Dana Farber Cancer Institute, Harvard Medical School, Boston, Massachusetts 02215.

products appear to be mediated by a single receptor (TcR). This receptor has been identified on murine and human T-cell clones using clone-specific (clonotypic) antibodies that bind to an 80- to 90-kDa glycoprotein containing two disulphide linked chains, α and β (Meuer *et al.*, 1983a–c; Haskins *et al.*, 1983; Samelson *et al.*, 1983). Similar molecules are present at the surface of cultured T-cell tumor lines (Allison *et al.*, 1982; Kappler *et al.*, 1983). They have been characterized as antigen receptors on the basis of clonal distribution, involvement in triggering antigen-specific T-cell functions, the presence of constant and variable peptides, and sequence correlation with T cell-specific, immunoglobulinlike rearranging genes (Meuer *et al.*, 1983b; Kappler *et al.*, 1983; Acuto *et al.*, 1984; Hedrick *et al.*, 1984; Sim *et al.*, 1984). The TCR is physically associated in human and murine T-cell membranes with a complex of proteins termed T3 that may serve to transduce an activating signal to the cell interior (reviewed in Terhorst, 1984; Terhorst and van den Elsen, 1985; Terhorst *et al.*, 1986; Weiss *et al.*, 1986).

2. Antigen Recognition by T-Cell Receptors after the Establishment of Cell–Cell Interactions

The cell–cell interactions that are involved in T-cell function can be studied in detail with the use of cytotoxic T lymphocytes (CTL). These cells play a critical role in the immune response to virus and parasite infections and in the destruction of tumor cells. The mechanism by which CTL lyse target cells can be separated into several stages. In the initial phase cell–cell contact between the CTL and its target cell is established. Next, "programming for lysis" takes place followed by the "lethal hit." Once the lethal hit has been delivered the presence of the CTL is not required for completion of the lysis of the target cell.

In order to distinguish the roles of the T-cell receptor/T3 complex from that of the accessory molecules, T4, T8, T11, and LFA 1 in more detail, we have used human HLA-A2 and HLA-B7 specific CTL clones and various target cells in a so-called "single cell assay" (Van de Rijn *et al.*, 1984; Spits *et al.*, 1986). This assay allows the measurement of the number of conjugates formed between killer cells and target cells and an estimate of the number of lysed target cells present in the conjugates. The results obtained in such studies clearly demonstrated that the cytotoxic reaction is initiated by nonspecific conjugate formation involving the accessory molecules.

In our model, the class I MHC antigens probably interact with T8 on the CTL. Since T8 has the ability to form homomultimers (Snow and Terhorst, 1983), its role could be to mobilize class I MHC antigens on the interface between CTL and target cells. We therefore speculate that the effector cell–target cell adhesion enables the T-cell receptors to enter the interface once a cluster of MHC class I molecules is present. T-cell receptors can then interact with class I antigens in a cooperative fashion. After the interaction between the T-cell receptor and its target antigen takes place, a cascade of reactions (Berke, 1980) leading to target cell lysis is initiated. If the relevant target antigen is not found the CTL detaches from the target cell.

It is clear from these observations (Spits *et al.*, 1986) that the requirements for a functional interaction between the T-cell receptor and its antigen are just beginning to be elucidated. It will be of interest to determine whether the rules governing antigen recognition by CTL also apply to the recognition of antigen by antigen-specific helper T cells, which generally see antigen in the context of class II MHC antigens.

3. The T-Cell Receptor–T3 Structure

3.1. Description of the T3 Antigen and Its Relationship to the T-Cell Receptor

The TcR-α and -β chains are associated with the T3 protein complex, which is defined on human lymphocytes by a group of monoclonal antibodies that recognize an antigen present on all peripheral blood T cells and 30–60% of thymocytes (Kung *et al.*, 1979; Reinherz *et al.*, 1980a,b; Hansen *et al.*, 1984). A physical interaction of T3 with TcR was initially suggested by the finding that incubation of T cells with antibodies directed against either structure leads to a disappearance or ''modulation'' of both from the T-cell surface together with a loss of antigen specific functions (Reinherz *et al.*, 1982; Meuer *et al.*, 1983d). When antibodies are removed from culture, TcR–T3 reexpression and normal function occur within 48 hr. Exposure of antigen-specific T-cell clones to high doses of antigenic peptides has been shown to induce T3 modulation and loss of function (Zanders *et al.*, 1983). Studies with ^{125}I surface-labeled cells indicate that modulation results in internalization of the T3 chains (Rinnooy Kan *et al.*, 1983). By immunoelectron microscopic analysis, the TcR–T3 chains and modulating antibody appear to be taken up into multivesicular bodies that subsequently fuse with lysozomes (Clevers *et al.*, 1986a).

When purified by immunoprecipitation from detergent lysates of ^{125}I surface-labeled human T cells, T3 contains three proteins (Borst *et al.*, 1982, 1983a,b; Kanellopoulos *et al.*, 1983). These have been designated T3-γ (25–28 kDa), T3-δ (20 kDa), and T3-ϵ (20 kDa) (see Table I).

T3-δ and T3-ϵ, which have approximately the same molecular mass (20kDa), are distinguishable on the basis of several criteria: (1) T3-δ, but not -ϵ, contains N-linked oligosaccharides and its molecular mass is reduced to 14 kDa following deglycosylation (Borst *et al.*, 1983b; Kanellopoulos *et al.*, 1983); (2) T3-ϵ is preferentially labeled with ^{125}I-iodo-naphthyl-azide (^{125}INA) or 3-(tri-

Table I. Properties of the Proteins of the TcR–T3 Complex [a]

A. Human T3–TCR complex

Chain	Molecular weight (kDa)	Endo-F sensitivity	Surface labeled with [125]I	Labeled with [3]H- and ^{35}S-amino acids
TCR-α	43-49	+	+	+
TCR-β	38-44	+	+	+
T3-γ	25-28	+	+	−
T3-δ	20	+	+	+
T3-ε	20	−	+	+
T3-p28	28	−	−	+

B. Murine TCR–T3 proteins

Chain	Molecular weight (kDa)	Endo-F sensitivity [b]	Surface labeled with [125]I	Labeled with [3]H- and ^{35}S-amino acids
TCR-α	40-50	+	+	+
TCR-β	40-50	+	+	+
T3-δ	28	+	+	+
T3-p25	25	−	+	+
T3-p21	21	+	+	+
T3-p17 (T3-ζ)	17	−	+	+
T3-p14	14	−	+	+

[a]References: Barker *et al.*, 1984; Caccia *et al.*, 1984; Collins *et al.*, 1985; Croce *et al.*, 1979; D'Eustachio *et al.*, 1980, 1981; Erlich *et al.*, 1983; Itakura *et al.*, 1971; Jones *et al.*, 1985; Klein and Schreffler, 1971; Krauz *et al.*, 1985; McBride *et al.*, 1982; Murré *et al.*, 1985; Swan *et al.*, 1979; van den Elsen *et al.*, 1985, 1986.
[b]Endo-F (Endo-β-N-acetylglucosaminidase F) sensitivity indicates the presence of N-linked carbohydrate.

fluoromethyl)-3-(m-[^{125}I]-iodophenyl) diazirine, two hydrophobic, membrane-partitioning reagents that form a reactive nitrene or carbene respectively upon ultraviolet irradiation (Borst *et al.*, 1983b; R. Malin and C. Terhorst, unpublished); (3) the [125]INA-labeled T3-ε chain is resolved as a single band in isoelectric focusing (IEF) analysis while biosynthetically or ^{125}I surface-labeled T3 contains proteins with multiple isoelectric points (pIs) (Borst *et al.*, 1983b); (4) the two chains have distinct N-terminal amino acid sequences and tryptic peptides (Borst *et al.*, 1984; Gold *et al.*, 1986a); (5) monoclonal antibodies raised against purified, denatured 20-kDa T3 can be shown by immunoblotting to react either with a heterogeneously charged 20-kDa glycoprotein (and with its 14-kDa deglycosylated polypeptide backbone) (T3-δ) or with a slightly more basic 20-kDa protein lacking N-linked carbohydrate (T3-ε) (Pessano *et al.*, 1985).

The 25- to 28-kDa T3-γ chain is coprecipitated with the δ and ε polypeptides from ^{125}I surface-labeled preparations regardless of the anti-T3 reagent used. It is resolved by SDS-polyacrylamide gel electrophoresis (SDS-PAGE) as a diffuse band of low intensity. On two dimensional (SDS-PAGE : IEF) gels, T3-γ appears as a relatively basic series of spots (Borst *et al.*, 1982). Its charge heterogeneity is reduced by the removal of terminal sialic acids or by cleavage of all N-linked sugars. The deglycosylated T3-γ chain has a molecular weight of 16 kDa. Unlike T3-δ, which contains high mannose N-linked oligosaccharides, T3-γ has only complex-type sugars.

An additional T3 protein, p28, is recovered only after short incubations with [35]S- or ^{3}H-amino acids and is rapidly lost with increasing labeling or chase times (Pettey et al., 1986). In the presence of monensin, an inhibitor of processing steps occurring in acidic intracellular compartments, T3-p28 is accumulated (Pettey et al., 1986). The protein is distinct from the 25- to 28-kDa T3-γ chain as determined by its sharp resolution on SDS-PAGE (T3-γ is diffuse), the absence of oligosaccharide side chains, its resistance to extraction with mild detergents (which effectively solubilize T3-γ) and its failure to incorporate ^{125}I in lactoperoxidase-catalyzed cell surface iodination (Pettey et al., 1986). Given its apparent absence from the cell surface and its transient association with T3, T3-p28 may play a role in the assembly and intracellular transport of the complex.

3.2. The Murine T-Cell Receptor–T3 Complex

Recently several investigators have identified a group of proteins that are associated with murine TcR. By treating ^{125}I surface-labeled murine lymphoma cells with a bifunctional cross-linker prior to solubilization, Allison and Lanier (1985) recovered two proteins (24 kDa and 29 kDa) that are specifically associated with TcR.

Using mild extraction conditions (cell solubilization with digitonin), a TcR-associated complex of four to five membrane associated proteins (14–28 kDa) has been identified (Table I; Oettgen et al., 1986). Like human T3, these include two glycosylated monomeric polypeptides (21 and 28 kDa) and a third chain (25 kDa), which is resistant to enzymes which remove N-linked oligosaccharides. In addition, the murine TcR-associated proteins include a series of dimers composed of 14- to 17-kDA subunits (T3-ζ) unlike any of the members of the human TCR–T3 complex (Oettgen et al., 1986). On the basis of immunoblotting experiments with a serum directed against a synthetic peptide corresponding to a section of the murine T3-δ cDNA sequence, the largest of the TcR-associated proteins (28 kDa) has been identified as the T3-δ chain (Oettgen et al., 1986). A very similar set of TcR-associated proteins has been described by Samelson et al. (1985b).

Proteins composed of 14- to 17-kDa subunits have not yet been observed in association with human T3 or TcR. The family of dimers specifically coprecipitated from lysates of murine T-cell hybridomas with anti-TCR reagents (T3-ζ), therefore, represents a new member of the group of TCR–T3 complex polypeptides. As the cells used in investigations of murine T3 were hybrids derived from fusions with the AKR thymoma, BW5147, it was possible that the dimers were contributed by this line. Preliminary analyses of anti-TcR-allotype immunoprecipitates prepared from digitonin extracts of normal murine splenic non-adherant cells indicate that TcR from this source is also associated with dimeric T3-ζ molecules (H. Oettgen and C. Terhorst, unpublished data). Whether they are integral membrane proteins, external membrane-associated proteins or perhaps ligands for a TcR–T3 acceptor site remains to be investigated. Their appearance as a cluster on two-dimensional nonreducing/reducing gels suggests that the 14- and 17-kDa subunits are related. It is possible that the series of dimers is derived by proteolytic processing of a 32-kDa parent homodimer. The presence of these proteins in the TcR–T3 complex of murine T cell lines, which are capable of functional responses to specific antigens, indicates that they may play a role in the events of antigen recognition and initiation of subsequent intracellular processes.

3.3. The T3-δ Chain

Several conclusions can be drawn from the amino acid sequence, which was derived from the nucleotide sequence of the T3-δ cDNA (Van den Elsen *et al.*, 1984). The T3-δ protein consists of three domains, an extracellular domain (residue 1–79), a transmembrane segment (80–106), and an intracellular domain (107–150) (Fig. 1). The transmembrane segment consists of a stretch of 27 predominantly hydrophobic amino acids (residues 80–106) that is interrupted by an aspartic acid (residue 90). Asp 90 could form a salt bridge with the Lys residue in the transmembrane segment of the T3–T-cell receptor α or β chains. This type of bond would be pronounced due to the low dielectric constant of the hydrophobic environment. The extracellular domain contains two N-linked oligosaccharides.

3.4. The Amino Acid Sequence of T3-ε
Derived from the cDNA Nucleotide Sequence

Using a bacteriophage expression vector (lambda gt11) and a rabbit anti-T3-δ and -T3-ε serum, we recently isolated a cDNA that codes for the T3-ε protein, (Gold *et al.*, 1986a). Proof of the identity of this clone as T3-ε was derived from:
1. Reactivity of the β-galactosidase fusion protein with a monoclonal anti-T3-ε reagent SP-6 (Pessano *et al.*, 1985),
2. Matching of a partial amino acid sequence of T3-ε with the protein sequence predicted from the nucleotide sequence of the cDNA (Gold *et al.*, 1986a).

A hydropathicity plot of the entire coding segment of the T3-ε cDNA predicted the presence of a putative N-terminal signal peptide containing an 11-amino acid hydrophobic core. This signal peptide is followed by a hydrophilic

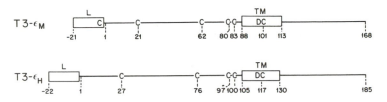

Figure 1. General features of the organization of the human T3δ and ε-chains. L, leader sequence; TM, transmembrane region; C, cysteine; N, asparagine; CHO, N-linked glycolylation sites. The shaded areas in T3-ε denote an uncharged region. Numbering refers to the relative position of residues in the mature protein.

domain 104 amino acids in length. This stretch begins with the amino terminus of the mature T3-ε protein. A 26-amino-acid-long hydrophobic segment, beginning at position 105 delineated a putative hydrophobic transmembrane region. The segment between residues 131 and 180 is hydrophilic. The absence of potential sites for N-linked glycosylation in the translated amino acid sequence was consistent with our observation that the T3-ε contains no N-linked oligosaccharides (Borst *et al.*, 1984).

Although no strong over all homology exists between T3-δ and T3-ε an interesting feature is noteworthy. As was observed in the case of the T3-δ chain, the T3-ε protein contains an Asp residue in the transmembrane region at position 115. There is in T3-ε a stretch of 23 amino acids (83–115); nine of which are shared between T3-δ and T3-ε, while the remaining 14 residues contained mostly conservative changes. In addition, five of the six charged residues as well as the two Cys residues in that area are conserved in both the human and murine T3-δ chains. In particular, the negatively charged Asp residue in the transmembrane region is of great interest due to the association of the T3 proteins with the T-cell receptor heterodimer. Both the α and β chains of the T-cell receptor contain a Lys in their respective transmembrane regions. The negatively charged Asp residues in both T3-δ and T3-ε could form salt bridges with these charged residues of the heterodimer α and β chains.

Based on the homology between T3-δ and T3-ε, we postulate that the region between T3-δ and T3-ε residues 105 and 130 is a transmembrane segment oriented in a fashion similar to the T3-δ chain; i.e., the N-terminus domain is extracellular and the C-terminal domain is cytoplasmic.

Upon comparison with other known protein sequences some homology could be detected with members of the immunoglobulin supergene family. Although this homology is restricted to areas near Cys residues and large gaps in the T3-δ and T3-ε extracellular domains need to be introduced (Fig. 2), it seems plausible that the T3-proteins are distantly related to the immunoglobulin supergene family.

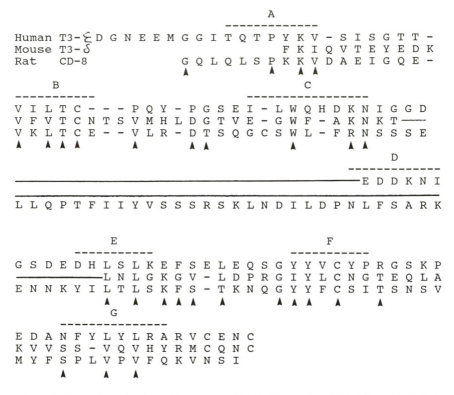

Figure 2. Comparison of amino acid sequences of T3-δ, T3-ε, and rat CD-8. The region indicated by A–F represents β-pleated sheets that are found in members of the immunoglobulin supergene family.

3.5. Physical Association of the T-Cell Receptor and T3 Proteins

Direct structural analysis has provided evidence for the association of TcR–T3 as a complex. Immunoprecipitates made from conventional cell lysates (prepared using the detergents Nonidet P-40 (NP40) or Triton X-100) with anti-T3 antibodies contain predominantly T3-δ, -ε, and -γ chains with minimal amounts of associated TcR-α and -β (Borst *et al.*, 1982, 1983a; Kanellopoulos *et al.*, 1983; Ledbetter *et al.*, 1981; Meuer *et al.*, 1983c; Oettgen *et al.*, 1984). Anti-TcR immunoprecipitates from such lysates are rarely observed to contain any of the T3 chains (Oettgen *et al.*, 1984; Spits *et al.*, 1985a).

Under some conditions, the complex can be extracted intact and all five 125-I-labeled chains are recovered regardless of the T3 or TcR specificity of the antibody. T3 preparations from the cells of a Sezary syndrome patient, for example, have been shown to contain a significant amount of TcR-α and -β (Oettgen *et al.*, 1984). Stabilization of the complex with bifunctional crosslinking reagents prior to solubilization has also permitted the isolation of intact TcR–T3

(Brenner *et al.*, 1985). These crosslinking studies have further demonstrated a spatial relationship between T3-γ and TCR-β chains in the plasma membrane. Extraction of labeled cells with the mild detergent, digitonin, preserves the TcR–T3 association in solution such that the complete complex is precipitated using either anti-T3 or anti-TcR antibodies (Oettgen *et al.*, 1986).

3.6. Coordinate Cell-Surface Expression of the T-Cell Receptor–T3 Complex

Cell surface expression of the TcR-α and -β proteins and T3-γ, -δ, and -ε chains requires the presence of all the components. Mutants of the human T leukemic cell line, Jurkat, selected for absence of T3 antigen invariably have concomitant loss of the TcR epitope (Weiss and Stobo, 1984). Conversely, cells selected for the loss of TcR are also T3-negative. Biosynthetically labeled (TcR–T3)⁻ cells contain intracellular T3 proteins similar to those observed in the wild-type cells (Ohashi *et al.*, 1985). Three of the seven mutant clones analyzed have diminished levels of TcR-α transcripts. Transfection of one of these clones by protoplast fusion with a plasmid containing a α-chain cDNA sequence and a drug marker gave rise to cells expressing the entire TcR–T3 complex at the surface (Ohashi *et al.*, 1985). This finding supports the hypothesis that all TcR–T3 proteins must be present prior to aseembly of the functional complex.

4. Organization of the Genes Encoding the T3 Proteins

4.1. The T3-δ Gene

Genomic DNA clones containing the gene coding for the 20-kDa T3 glycoprotein of the T-cell receptor–T3 complex (T3-δ chain) of both human and mouse were isolated and characterized (Van den Elsen *et al.*, 1986). The human T3-δ gene is approximately 4-kb long and contains five exons: a 151-bp exon containing the 5′ untranslated and the coding sequences of the signal peptide, one exon of 219-bp that contains most of the extracellular segment of the T3-δ chain, one 130-bp-long exon coding mainly for the transmembrane portion of the molecule and two exons of 44-bp and 156-bp respectively encoding the cytoplasmic domain and 3′ untranslated region of the T3-δ chain. The murine T3-δ gene, which has a similar organization spans 5-kb, because the first intron is approximately 1-kb larger than in the human gene.

Two major mRNA initiation sites within a small area approximately 100 nucleotides 5′ of the AUG codon were identified by S1 nuclease analysis and primer extension studies (Van den Elsen *et al.*, 1986). The remarkably high level of conservation of nucleotide sequences in this region suggest that this segment may be important for the regulation of T-cell specific transcription of the T3-δ gene. The T3-δ gene does not contain the TATA box found in many eukaryotic promoters.

4.2. The T3-ε Gene

The human and murine T3-ε genes have been isolated (T. Wileman, H. Clevers, and C. Terhorst, unpublished). They are located on a 9-kb segment in the human and on a 6-kb segment in the mouse.

4.3. Chromosomal Locations of the T3-δ and -ε Genes

T3-δ and T3-ε are encoded on the distal portion of the long arm of human chromosome 11 (11q23–11qter) or mouse chromosome 9 (See Table II; Van den Elsen, 1985b; Gold *et al.*, 1986b). Thus, these T3 chains are not linked to MHC, TcR, or Ig genes in either species.

The human gene homologue of murine Thy-1 also maps to the long arm of chromosome 11 (Seki *et al.*, 1985) but its product does not appear to be part of the T3 complex. In addition, loci controlling at least 12 other cell-surface differentiation antigens have been assigned to chromosome 11, more than are known to map to any other chromosome (reviewed in Rettig *et al.*, 1985). The 11q23–11qter region contains a cluster of breakpoints found in various tumors (Mitelman, 1984; Yunis, 1983. In Ewing's sarcomas, a specific translocation [t(11;22) (q24;q12)] is observed that results in the transfer of the c-*sis* oncogene to this region (Aurias *et al.*, 1983; Turc-Carel *et al.*, 1983; van Kessel *et al.*, 1985). Hu-*ets*-1, one of two human genomic sequences homologous the avian retrovirus E26 oncogene, v-*ets*, also maps to 11q23–11qter (DeTaisne *et al.*, 1984). Chromosomal abnormalities of 11q23 in some leukemic and lymphomatous cells are associated with rearrangements and amplification of c-*ets* sequences (Rovigatti *et al.*, 1986). In chickens, an antipeptide serum has been used to

Table II. Chromosomal Localization of Genes Encoding TCR-T3, Thy-1, the Major Histocompatibility Complex, and Immunoglobulin[a]

Gene(s)	Human chromosome	Mouse chromosome
TCR-α	14	14
TCR-β	7	6
TCR-γ	7	13
T3-δ	11	9
T3-ε	11	9
MHC	6	17
Thy-1	11	9
Ig-H	14	12
Ig-κ	2	6
Ig-λ	22	16

[a]References as cited in text.

identify the c-*ets* product as a 54-kDa cytosolic protein present in small amounts in most tissues examined but expressed at high levels in thymocytes and bursal lymphocytes (Ghysdael *et al.*, 1986). Thus the T3-δ and -ε chain genes, and other antigens mapping to the long arm of chromosome 11, are contained in a region that may encode products that influence normal cellular differentiation and growth control.

5. Expression of the T3 Genes during Intrathymic Differentiation

T-cell receptor and T3 appearance at the cell surface in thymic ontogeny are coordinated and occur primarily in the most mature thymoctyes and in T-cell tumors with a similar phenotype (Acuto *et al.*, 1983; Royer *et al.*, 1985; Furley *et al.*, 1986). The sequence of human TcR-T3 gene rearrangement, transcription, and translation leading to surface expression has been analyzed using cDNA probes and specific antibodies. T3-δ, and -ε chain transcripts are detectable in the least mature thymocytes (T3−, T6−, T4−, T8−) and in corresponding T-cell tumors (Furley *et al.*, 1986). Their protein products are accumulated intracellularly in a perinuclear distribution. Rearrangement and high levels of transcription of the TcR-β chain are observed in T3−, T11+, T6+, T4+, T8+ thymocytes and in their putative progeny, the immunocompetent T11+, T3+, T4+, or T8+ T cells (Furley *et al.*, 1986). Based on Southern blot analyses of T-cell tumors, it appears that TcR-β rearrangements may occur even prior to the acquisition of T6, T4, and T8, but this may reflect an increased tendency towards malignant transformation of the cells undergoing rearrangement (Furley *et al.*, 1986). TcR-α transcription appears to be the final rate-limiting step as mRNA for this protein is observed only in thymocytes and tumors with surface TcR–T3.

The sequence of expression of murine TcR and T3 genes has been examined in subsets of lymphocytes from mature thymus and in fetal thymoctyes. A pattern of rearrangement and transcription of the TcR–T3 genes parallel to that described above for human thymoctyes has been observed. Cells of the dLy1+ phenotype obtained from mature thymus contain T3-δ transcripts (Samelson *et al.*, 1985a). In addition, they express TcR-β mRNA but no TcR-α message. Southern blots of DNA from hybrids derived from these cells by fusion with a thymoma indicate that eight out of twelve contain rearranged β-genes. In fetal murine thymus, β-chain mRNA is detectable as early as day 15 (when only dLy1+ cells are present) (Raulet *et al.*, 1985; Snodgrass *et al.*, 1985). Transcripts for the TcR-γ chain are also at high levels in d15 cells. Southern blot analyses of hybrids derived from fetal cells show that β rearrangements occur by d14 and reach a maximum frequency by d16 (Born *et al.*, 1985). The appearance of α chain transcripts occurs approximately two days later (d17), at about the same time when Ly1+, Ly2+, L3T4+ cells arise from their dLy1+ precursors. Day 17 is also the earliest point at which cell-surface TcR can be detected in these cells using a monoclonal anti-TcR-allotype reagent KJ16-133 (Roehm *et al.*, 1984). Prior to cell surface expression, the TcR-β chain epitope recognized by this reagent accumulates intracellularly in the perinuclear envelope as is also observed for human T3-δ and -ε in immature thymoctyes (Farr *et al.*, 1985).

These studies of thymocyte cell-surface phenotypes and TcR gene expression reveal several parallels between human and murine T-cell development. Rearrangements of TcR-β and transcription of the β chain and T3-δ/-ε occur at a relatively early stage in both species with TcR-α expression following later. Only cells that express both of these products appear to have surface TcR or T3. These mature thymocytes are also defined by their exclusive expression of either T4/L3T4 or T8/Ly2.

ACKNOWLEDGMENTS. This work was supported by grants from the National Institutes of Health, the American Cancer Society, and the Leukemia Society of America.

References

Acuto, O., Hussey, R.E., Fitzgerald, K.A., Protentis, J.P., Meuer, S.C., Schlossman, S.F., and Reinherz, E.L., 1983, The human T cell receptor: appearance in ontogeny and biochemical relationship of α and β subunits on IL-2 dependent clones and T cell tumors, *Cell* **34**:717.

Acuto, O., Fabbi, M., Smart, J., Poole C.B., Protentis, J., Royer, H.D., Schlossman, S.F., and Reinherz, E.L., 1984, Purification and NH$_2$-terminal amino acid sequencing of the β subunit of a human T-cell antigen receptor, *Proc. Nat'l. Acad. Sci. USA* **81**:3851.

Allison, J.P., McIntyre, B.W., and Bloch, D., 1982, Tumor specific antigen of a murine T-lymphoma defined with monoclonal antibody, *J. Immunol.* **129**:2293.

Allison, J.P., and Lanier, L.L., 1985, Identification of antigen receptor-associated structures on murine T cells, *Nature* **314**:107.

Aurias, A., Buffe, C., Dubounnet, J., and Marabrand, A., 1983, Chromosomal translocation in Ewing's sarcoma, *N. Eng. J. Med.* **309**:496.

Barker, P.E., Ruddle, F.H., Royer, H.D., Acuto, O., and Reinherz, E.L., 1984, Chromosomal location of human T-cell receptor gene Ti α, *Science* **226**:348.

Berke, G., 1980, Prog., *Allergy* **27**:69.

Born, W., Yague, J., Palmer, E., Kappler, J., and Marrack, P., 1985, Rearrangement of T-cell receptor Ā-chain genes during T-cell development, *Proc. Natl. Acad. Sci. USA* **82**:2925.

Borst, J., Prendiville, M.A., and Terhorst, C., 1982, Complexity of the human T lymphocyte-specific cell surface antigen T3, *J. Immunol.* **128**:1560.

Borst, J., Alexander, S., Elder, J., and Terhorst, C., 1983a, The T3 complex on human T lymphocytes involves four structurally distinct glycoproteins, *J. Biol. Chem.* **259**:5135.

Borst, J., Prendiville, M.A., and Terhorst C., 1983b, The T3 complex on human thymus-derived lymphocytes contains two different subunits of 20 kDa, *Eur. J. Immunol.* **13**:576

Borst, J., Colligan, J.E., Oettgen, H., Pessano, S., Malin, R., and Terhorst, C., 1984, the δ- and ε-chains of the human T3/T-cell receptor complex are distinct polypeptides, *Nature* **312**:455.

Brenner, M.B., Trowbridge, I.S., and Strominger, J.L., 1985, Cross-linking of human T cell receptor proteins: Association between the T cell idiotype β subunit and the T3 glycoprotein heavy subunit, *Cell* **40**:183.

Caccia, N., Bruns, G.A., Kirsch, I.R., Hollis, G.F., Bertness, V., and Mak, T.W., 1984a, T cell receptor α-chain genes are located on chromosome 14 at 14q11-14q12 in humans, *J. Exp. Med.* **161:**1255.

Clevers, H., Peters, P., Pettey, C., Ballieux, R.E., and van der Donk, H.A., 1986, Modulation of T3/antigen receptor complex of the human T lymphocyte is an endocytotic process, *J. Immunol.* (in press).

Collins, M.K.L., Goodfellow, P.N., Spurr, N.K., Solomon, E., Tanigawa, G., Tonegawa, S., and Owen, M.J., 1985, The human T-cell receptor α-chain maps to chromosome 14, *Nature* **314:**273.

Croce, C.M., Shander, M., Martinis, J., Cicurel, L., D'Ancona, G.G., Dolby, T.W., and Koprowski, H., 1979, Chromosomal location of the genes for human immunoglobulin heavy chains, *Proc. Natl. Acad. Sci. USA* **76:**3416.

DeTaisne, C., Gegonne, A., Stehelin, D., Bernhein, A., and Berger, R., 1984, Chromosomal location of the human proto-oncogene c-*ets*, *Nature (Lond.)* **310:**581.

D'Eustachio, P., Pravtcheva, D., Marcu, K., and Ruddle, F.H., 1980, Chromosomal location of the structural gene cluster encoding murine immunoglobulin heavy chains, *J. Exp. Med.* **151:**1545.

D'Eustachio, P., Bothwell, A.L.M., Takaro, T.K., Baltimore, D., and Ruddle, F.H., 1981, Chromosomal location of structural genes encoding murine immunoglobulin light chains, *J. Exp. Med.* **153:**793.

Erlich, H.A., Stetler, D., Saiki, R., Gladstone, P., and Pious, D., 1983, Mapping of the genes encoding the HLA-Dr α chain and the HLA-related antigens to a chromosome 6 deletion by using genomic blotting, *Proc. Natl. Acad. Sci. USA* **80:**2300.

Farr, A.G., Anderson, S.K., Marrack, P., and Kappler, J., 1985, Expression of antigen-specific, major histocompatibility complex-restricted receptors by cortical and medullary thymocytes in situ, *Cell* **43:**543.

Furley, A.J., Mizutami, S., Weilbaecher, K., Dhaliwal, H.S., Ford, A.M., Chan, L.C., Molgaard, H.V., Toyonaga, B., Mak, T., van den Elsen, P., Gold, D., Terhorst, C., and Greaves, M.F., 1986, Developmentally regulated rearrangement and expression of genes encoding the T cell receptor-T3 complex, *Cell* (in press).

Ghysdael, J., Gegonne, A., Pognonec, P., Leprince, D., and Stehelin, D., 1986, Identification and preferential expression in thymic and bursal lymphocytes of a c-*ets* oncogene-encoded M_r 54,000 cytoplasmic protein, *Proc. Natl. Acad. Sci. USA* **83:**1714.

Gold, D.P., Puck, J.M., Pettey, C.L., Cho, M., Colligan, J., Woody, J.N., and Terhorst, C., 1986a, Isolation of cDNA clones encoding the 20 kD non-glycosylated polypeptide chain (T3-ε) of the human T cell receptor/T3 complex, *Nature* **321:**431.

Gold, D.P., van Dongen, J.J.M., Morton, C.C., Bruns, G., van den Elsen, P.E., van Kessel, G., and Terhorst, C., 1987, The gene encoding the ε subunit of the T3/T cell receptor complex maps to chromosome 11 in humans and to chromosome 9 in mice, *Proc. Natl. Acad. Sci. USA* **84:**1664.

Hansen, J.A., Martin, P.J., Beatty, P.G., Clark, E.A., and Ledbetter, J.A., 1984, Human T lymphocyte cell surface molecules defined by the workshop monoclonal antibodies ("T cell protocol"), in: *Leukocyte Typing, Human Leukocyte Differentiation Antigens Detected by Monoclonal Antibodies* (A. Bernard, L. Boumsell, J. Dausset, C. Milstein, and S.F. Schlossman, eds.), Berlin, Springer-Verlag, p. 195.

Haskins, K., Kubo, R., White, J., Pigeon, M., Kappler, J., and Marrack, P., 1983, The major histocompatibility complex-restricted antigen receptor on T cells. I. Isolation with a monoclonal antibody, *J. Exp. Med.* **157:**1149–1169.

Hedrick, S.M., Cohen, D.I., Nielsen, E.A., and Davis, M., 1984, Isolation of cDNA clones encoding T cell-specific membrane-associated proteins, *Nature* **308:**149.

Itakura, K., Hutton, J.J., Boyse, E.A., and Old, L.J., 1971, Linkage between mouse antigen loci, *Nature* **230:**126.

Johnson, P., Barclay, N., and Williams, A., 1985, Purification, chain separation, and sequence of the MRC OX-8 antigen, a marker of rat cytotoxic T lymphocytes, *EMBO J.* **4:**2539.

Jones, C., Morse, H.G., Kao, F.T., Carbone, A., and Palmer, E., 1985, Human T-cell receptor α-chain genes: location on chromosome 14, *Science* **228:**83.

Kanellopoulos, J.M., Wigglesworth, N.M., Owen, M.J., and Crumpton, M.J., 1983, Biosynthesis and molecular nature of the T3 antigen of human T lymphocytes, *EMBO J.* **2:**1807.

Kappler, J., Kubo, R., Haskins, K., Hannum, C., Marrack, P., Pigeon, M., McIntyre, B., Allison, J., and Trowbridge, I., 1983, The major histocompatibility complex-restricted antigen receptor on T cells in mouse and man: identification of constant and variable peptides, *Cell* **35:**295.

Klein, J., and Shreffler, D.C., 1971, the H-2 model for the major histocompatibility system, *Transpl. Rev.* **6:**3.

Kranz, D.M., Saito, H., Disteche, C.M., Swisshelm, K., Pravtcheva, D., Ruddle, F.H., Eisen, H.N., and Tonegawa, S., 1985. Chromosomal locations of the murine T-cell receptor α-chain gene and the T-cell γ-gene. *Science* **227:**941.

Kung, P.C., Goldstein, G., Reinherz, E., and Schlossman, S.F., 1979, Monoclonal antibodies defining distinctive human T cell surface antigens, *Science* **206:**347.

Ledbetter, J.A., Evans, R.L., Lipinski, M., Cunningham-Rundles, C., Good, R.A., and Herzenberg, L.A., 1981, Evolutionary conservation of surface molecules that distinguish T lymphocyte helper/inducer and cytotoxic/suppressor subpopulations in mouse and man, *J. Exp. Med.* **153:**310.

McBride, O.W., Hieter, P.A., Hollis, G.F., Swan, D., Otey, M.C., and Leder, P., 1982, Chromosomal location of human kappa and lambda immunoglobulin light chain constant region genes, *J. Exp. Med.* **155:**1480.

Meuer, S.C., Acuto, O., Hussey, R.E., Hodgdon, J.C., Fitzgerald, K.A., Schlossman, S.F., and Reinherz, E.L., 1983a, Evidence for the T3-associated 90KD heterodimer as the T cell antigen receptor, *Nature* **303:**808.

Meuer, S.C., Hodgdon, J.C., Hussey, R.E., Protentis, J.P., Schlossman, S.F., and Reinherz, E.L., 1983b, Antigenlike effects of monoclonal antibodies directed at receptors on human T cell clones, *J. Exp. Med.* **158:**988.

Meuer, S.C., Cooper, D.A., Hodgdon, J.C., Hussey, R.E., Fitzgerald, K.A., Schlossman, S.F., and Reinherz, E.L., 1983c, Identification of the antigen/MHC-receptor on human inducer T lymphocytes, *Science* **222:**1239.

Meuer, S.C., Fitzgerald, K.A., Hussey, R.E., Hodgdon, J.C., Schlossman, S.F., and Reinherz, E.L., 1983d, Clonotypic structures involved in antigen-specific human T cell function: relationship to the T3 molecular complex, *J. Exp. Med.* **157:**705.

Mitelman, F., 1984, Restricted number of chromosomal regions implicated in aetiology of human cancer and leukaemia, *Nature* **310:**325.

Murre, C., Waldmann, R.A., Morton, C.C., Bongiovanni, D.F., Waldman, T.A., Shows, T.B., and Seidman, J.G., 1985, Human γ-chain genes are rearranged in leukaemic T cells and map to the short arm of chromosome 7, *Nature* **316:**549.

Oettgen, H.C., Kappler, J., Tax, W.J.M., and Terhorst, C., 1984, Characterization of the two heavy chains of the T3 complex on the surface of human T lymphocytes, *J. Biol. Chem.* **259:**12039.

Oettgen, H.C., Pettey, C.L., Maloy, W.L., and Terhorst, C., 1986, A T3-like protein complex associated with the antigen receptor on murine T cells, *Nature* **320:**272.

Ohashi, P.S., Mak, T.W., van den Elsen, P., Yanagi, Y., Yoshikai, Y., Calman, A., Terhorst, C., Stobo, J.D., and Weiss, A., 1985, Reconstitution of an active surface T3/T-cell antigen receptor by DNA transfer, *Nature* **316:**606.

Pessano, S., Oettgen, H., Bhan, A.K., and Terhorst, C., 1985, The T3/T cell receptor complex: antigenic distinction between the two 20-kd T3 (T3-δ and T3-ε) subunits, *EMBO J.* **4:**337.

Pettey, C., Alarcon, B., Malin, R., Weinberg, K., and Terhorst, C., 1987, T3-p28 is a protein associated with the δ and ε chains of the T cell receptor–T3 antigen complex during biosynthesis, *J. Biol. Chem.* **262:**4854.

Raulet, D.H., Germain, R.D., Saito, H., and Tonegawa, S., 1985, Developmental regulation of T-cell receptor gene expression, *Nature* **314:**103.

Reinherz, E.L., Kung, P.C., Goldstein, G., Levey, R.H., and Schlossman, S.F., 1980a, Discrete stages of human intrathymic differentiation: Analysis of normal thymocytes and leukemic lymphoblasts of T-cell lineage, *Proc. Natl. Acad. USA* **77:**1588.

Reinherz, E.L., Hussey, R.E., and Schlossman, S.F., 1980b, A monoclonal antibody blocking human T cell function, *Eur. J. Immunol.* **10:**758.

Reinherz, E.L., Meuer, S.C., Fitzgerald, K.A., Hussey, R.E., and Schlossman, S.F., 1982, Antigen recognition by human T lymphocytes is linked to surface expression of the T3 molecular complex, *Cell* **30:**735.

Rettig, W.J., Dracopoli, N.C., Chesa, P.G., Spengler, B.A., Beresford, H.R., Davies, P., Biedler, J.L., and Old, L.J., 1985, Role of human chromosome 11 in determining surface antigenic phenotype of normal and malignant cells, *J. Exp. Med.* **162:**1603.

Rinnooy Kan, E.A., Wang, C.Y., Wang, L.C., and Evans, R.L., 1983, Noncovalently bonded subunits of 22 and 28 kD are rapidly internalized by T cells reacted with anti-Leu-4 antibody, *J. Immunol.* **131:**536.

Roehm, N., Herron, L., Cambier, J., DiGuisto, D., Haskins, K., Kappler, J., and Marrack, P., 1984, The major histocompatibility complex-restricted antigen receptor on T cells: distribution on thymus and peripheral T cells, *Cell* **38:**577.

Rovigatti, R., Watson, D.K., and Yunis, J.J., 1986, Amplification and rearrangement of Hu-*ets*-1 in leukemia and lymphoma with involvement of 11q23, *Science* **232:**398.

Royer, H.D., Ramarli, D., Acuto, O., Campen, T.J., and Reinherz, E.L., 1985, Genes encoding the T-cell receptor α and β subunits are transcribed in an ordered manner during intrathymic ontogeny, *Proc. Natl. Acad. Sci. USA* **82:**5510.

Samelson, L.E., Germain, R.N., and Schwartz, R.H., 1983, Monoclonal antibodies against the antigen receptor on a cloned T-cell hybrid, *Proc. Natl. Acad. USA* **80:**6972.

Samelson, L.E., Lindsten, T., Fowlkes, B.J., van den Elsen, P., Terhorst, C., Davis, M.M., Germain, R.N., and Schwartz, R.H., 1985a, Expression of genes of the T-cell antigen receptor complex in precursor thymocytes, *Nature* **315:**765.

Samelson, L.E., Harford, J.B., and Klausner, R.D., 1985b, Identification of the components of the murine T cell antigen receptor complex. *Cell* **43:**223.

Seki, T., Spurr, N., Obata, F., Goyert, S., Goodfellow, P., and Silver, J., 1985, The human Thy-1 gene: structure and chromosomal location, *Proc. Natl. Acad. Sci. USA* **82:**6657.

Sim, G.K., Yague, J., Nelson, J., Marrack, P., Palmer, E., Augustin, A., and Kappler, J., 1984, Primary structure of human T-cell receptor α-chain, *Nature* **312:**771.

Snodgrass, H.R., Dembic, Z., Steinmetz, M., and von Boehmer, H., 1985, Expression of T-cell antigen receptor genes during fetal development in the thymus, *Nature* **315:**232.

Snow, P., and Terhorst, C., 1983, The T8 antigen is a multimeric complex of two distinct subunits of human thymocytes, but consists of homomultimeric forms on peripheral blood T lymphocytes, *J. Biol. Chem.* **258:**14675.

Spits, H., Borst, J., Tax, W., Capel, P.J.A., Terhorst, C., and de Vries, J., 1985, Characteristics of a monoclonal antibody (WT-31) that recognizes a common epitope on the human T cell receptor for antigen, *J. Immunol.* **135:**1922.

Spits, H., Van Schooten, W., Keizer, H., Van Seventer, G., Van de Rijn, Terhorst, C., and De Vries, J., 1986, Alloantigen recognition is preceded by nonspecific adhesion of cytotoxic T cells and target cells, *Science* **232:**403.

Swan, D., D'Eustachio, P., Leinwand, L., Seidman, J., Keithley, D., and Ruddle, F.H., 1979, Chromosomal assignment of the mouse kappa light chain genes, *Proc. Natl. Acad. Sci. USA* **76:**2735.

Terhorst, C., van Agthoven, A., LeClair, K., Snow, P., Reinherz, E.L., and Schlossman, S.F., 1981, Biochemical studies of the human thymocyte cell surface antigens T6, T9, and T10, *Cell* **23:**771.

Terhorst, C., 1984, Cell surface structures involved in human T lymphocyte specific functions:

analysis with monoclonal antibodies, in: *Receptors and Recognition,* (M.F. Greaves, ed.) p. 133, Chapman and Hall, London.

Terhorst, C., and van den Elsen, P., 1985, The T-cell receptor/T3 complex, in: *The Year in Immunology 1984–1985,* (J.M. Cruse and R.E. Lewis, eds.) p. 62, S. Karger, Basel.

Terhorst, C., DeVries, J., Georgopoulos, K., Gold, D., Oettgen, H., Pettey, C., Spits, H., Ucker, D., van den Elsen, P., and Wileman, T., 1986 The T-cell receptor/T3 complex, in: *The Year in Immunology 1985–1986* (J.M. Cruse and R.E. Lewis, Jr., eds.), S. Karger, Basel, p. 245.

Turc-Carel, C., Philip, I., Berger, M.-P., Philip, T., and Lenoir, G., 1983, Chromosomal translocation in Ewing's sarcoma, *N. Engl. J. Med.* **309:**497.

van den Elsen, P., Shepley, B.-A., Borst, J., Colligan, J., Markham, A.F., Orkin, S., and Terhorst, C., 1984, Isolation of cDNA clones encoding the 20K T3 glycoprotein of the human T-cell receptor complex, *Nature* **312:**413.

van den Elsen, P., Shepley, B.-A., Cho, M., and Terhorst, C., 1985a, Isolation and characterization of a cDNA clone encoding the murine homologue of the human 20K T3/T-cell receptor glycoprotein, *Nature* **314:**542.

van den Elsen, P., Bruns, G., Gerhard, D.S., Pravtcheva, D., Jones, C., Housman, D., Ruddle, F.A., Orkin, S.F., and Terhorst, C., 1985b, Assignment of the gene coding for the T3-δ subunit of the T3-T-cell receptor complex to the long arm of human chromosome 11 and to mouse chromosome 9, *Proc. Natl. Acad. Sci. USA* **82:**2920.

van den Elsen, P., Georgopoulos, K., Shepley, B.-A., Orkin, S., and Terhorst C., 1986, Exon/Intron organization of the genes coding for the δ-chains of the human and murine T cell receptor/T3 complex, *Proc. Natl. Acad. Sci. USA* **83:**2944.

van de Rijn, M., Bernabeu, C., Royer-Pokora, B., Seidman, J., de Vries, J. and Terhorst, C., 1984, Recognition of HLA-A2 by cytotoxic T-lymphocytes after DNA transfer into human and murine cells, *Science* **226:**1083.

van Kessel, G., Turc-Carel, C., de Klein, A., Grosveld, G., Lenoir, G., and Bootsma, D., 1985, Translocation of oncogene c-*sis* from chromosome 22 to chromosome 11 in a Ewing's sarcoma-derived cell line, *Mol. Cell. Biol.* **5:**427.

Weiss, A., and Stobo, J.D., 1984, Requirement for the coexpression of T3 and the T cell antigen receptor on a malignant human T cell line, *J. Exp. Med.* **160:**1284.

Weiss, A., Imboden, J., Hardy, K., Manger, B., Terhorst, C., and Stobo, J., 1986, The role of the T3/antigen receptor complex in T-cell activation, *Annu. Rev. Immunol.* **4:**593.

Yunis, J.J., 1983, The chromosomal basis of human neoplasia, *Science* **221:**227.

Zanders, E.D., Lamb, J.R., Feldman, M., Green, N., and Beverley, P.C.L., 1983, Tolerance of T-cell clones is associated with membrane antigen changes, *Nature* **303:**625.

7

Role of the T3/T-Cell Antigen Receptor Complex in T-Cell Activation

BERNARD MANGER, JOHN IMBODEN, and ARTHUR WEISS

1. Introduction

Interactions that occur at the interface of the plasma membranes of T cells and antigen-presenting cells (or target cells) initiate events that culminate in T-cell activation. Activation may be manifested by the production of lymphokines, appearance of new cell surface antigens (such as the IL-2 receptor) and, ultimately, in T-cell proliferation. The T-cell antigen receptor (Ti) plays a central role in this interaction and has two functions: First, it must subserve a cognitive function, recognizing antigen in the context of the major histocompatibility complex (MHC). Secondly, it must convert such a recognition event into a transmembrane signal which can initiate T-cell activation. This chapter will focus on this second role of Ti in initiating human T-cell activation.

2. The T3/Ti Complex

Monoclonal antibodies (MAb) reactive with T cell clone-specific determinants were first used to identify the protein structure of Ti (Allison *et al.*, 1982; Meuer *et al.*, 1983a; Haskins *et al.*, 1983; Kaye *et al.*, 1983; Samelson *et al.*,

BERNARD MANGER, JOHN IMBODEN, and ARTHUR WEISS ● Howard Hughes Medical Institute and Department of Medicine, University of California, San Francisco, San Francisco, California 94143. *Present address for B. M.:* Institute of Clinical Immunology, Medical School Erlangen, D-8520 Erlangen, Federal Republic of Germany. *Present address for J. I.:* Immunology/ Arthritis Section, Veterans Administration Medical Center, San Francisco, California 94121.

1983; Weiss and Stobo, 1984). Ti consists of an 80- to 90-kD heterodimer, composed of an acidic α chain and a more basic β chain. On human cells, this heterodimer is noncovalently associated with three nonpolymorphic noncovalently linked peptides of 20–28 kD, termed T3 (Borst *et al.*, 1983). A molecular complex similar to T3 has been identified in association with the murine Ti, but appears to be more complex (Allison and Lanier, 1985; Samelson *et al.*, 1985; Samelson *et al.*, 1986).

The association between Ti and T3 is poorly understood. Their association was first demonstrated in comodulation studies in which T-cell clones exposed to anti-Ti expressed a markedly diminished amount of T3 on the cell surface (Meuer *et al.*, 1983a). The physical interaction between Ti and T3 has also been demonstrated in coimmunoprecipitation studies (Reinherz *et al.*, 1983) as well as chemical crosslinking studies (Brenner *et al.*, 1985). Studies in our laboratory aimed at studying the relative function of Ti and T3 revealed that cell-surface expression of T3 depends upon its association with the Ti heterodimer. In these studies, mutants of Jurkat, a human T-cell line, were selected for their failure to express either T3 or Ti on the cell surface. In every instance, however, these mutants were found to have concomitantly lost Ti and T3 from the plasma membrane (Weiss and Stobo, 1984). Characterization of these mutants revealed that T3 could be detected intracellularly, but at least some mutants failed to contain full-length Ti β-chain transcripts (Ohashi *et al.*, 1985). This suggested that T3 was trapped intracellularly due to the lack of an associated Ti heterodimer. Indeed, this was shown to be the case, as reconstitution of a Ti β-chain-deficient mutant by gene transfer resulted in not only the reexpression of Ti but also T3 (Ohashi *et al.*, 1985). Thus, the association of T3 and Ti would appear to be an obligate requirement for the cell-surface expression of T3. Detailed analysis of the interacting components are not available. It has been proposed that, based on conserved lysine and arginine residues in the transmembrane domains of the Ti α and β chains (Yanagi *et al.*, 1984; Yanagi *et al.*, 1985) and aspartic and glutamic acid residues in the T3 δ-, ε, and γ-chain transmembrane domains (van den Elsen *et al.*, 1984; Gold *et al.*, 1986; Krissansen *et al.*, 1986), interactions between these charged amino acids may be crucial in the association between T3 and Ti. Whereas the function of the Ti heterodimer must involve recognitive events, the function of T3 in this multimeric structure is not clear. Evidence, as discussed below, would suggest it may be important in transmembrane signaling events.

3. T3/Ti Antibodies Provide One Stimulus Required for Human T-Cell Activation

The role of the T3/Ti complex in T-cell activation has been clarified through the use of MAb reactive with this structure. Thus, by utilizing such

MAb as probes, which can function as agonists or antagonists, complex interactions occurring at the interface between the T cell and antigen-presenting cell can be mimicked. For instance, early studies demonstrated that anti-T3 MAb were mitogenic for peripheral T cells (Van Wauwe *et al.*, 1980). Given the close association between T3 and the Ti heterodimer, the agonist properties of T3 MAb would suggest that it may mimic antigen interacting with the Ti heterodimer on a polyclonal level. Further analysis has revealed, however, that simple binding of anti-T3 to the surface of the T cell is insufficient to initiate activation. Mitogenic responses depended upon the presence of accessory cells (AC) contained within the adherent cell population of peripheral blood mononuclear leukocytes (PBL) (Van Wauwe and Goossens, 1981; Kaneoka *et al.*, 1983). Other studies have suggested that in some cases, AC function can be substituted for by the phorbol ester, phorbol myristate acetate (PMA) (Rosenstreich and Mizel, 1979; Farrar *et al.*, 1980). Therefore, the ability of anti-T3 to stimulate nonadherent PBL in the presence of PMA has been investigated (Table I; Hara and Fu, 1985). Indeed, PMA could reconstitute the anti-T3 induced mitogenic response of adherent cell depleted PBL (Table I). This observation suggested that at least two stimuli were required for T-cell activation.

Since PBL represent a complex cell population and the resultant mitogenic response occurs a relatively long time following ligand binding, it was desirable to examine an earlier response of a homogeneous cell population. Therefore, we examined the ability of anti-T3 MAb, with or without PMA, to induce the T-cell leukemic line, Jurkat, to produce interleukin-2 (IL-2) or γ interferon (IFN). Anti-T3 induced Jurkat to produce IL-2 and IFN only if added together with PMA (Weiss *et al.*, 1984a,b,c). Similarly, an anti-Ti MAb could also induce Jurkat to produce IL-2, but only if added together with PMA (Weiss and Stobo, 1984). Whereas, anti-T3 and anti-Ti MAb had similar agonist properties in the activation of Jurkat, individual MAb reactive with other cell-surface molecules

Table I. Two Stimuli Are Required to Activate Peripheral Blood T Cells with OKT3

Responding cells[a]	Stimulus	−PMA c.p.m.[b]	+PMA c.p.m.[b]
Unfractionated PBL	None	3,446	2,230
	OKT3	92,657	205,563
Nonadherent PBL	None	4,286	3,478
	OKT3	6,148	155,757

[a]Unfractionated or nonadherent PBL were incubated at 10^6 cells/ml with the indicated stimuli (OKT3 1:1000 ascites dilution, PMA 0.5 ng/ml) for 72 hr.
[b]Figures represent c.p.m. [³H]-thymidine uptake (mean values of triplicates of a representative experiments; standard deviation less than 15%).

expressed on Jurkat failed to activate it to produce IL-2 (Fig. 1) (Weiss *et al.*, 1984c). Thus, these results would support the notion that two stimuli acting together upon the same cell are required for T-cell activation. One stimulus involves perturbation of the T3/Ti complex, regardless of whether the ligand interacts with Ti or T3. The other stimulus is normally provided by AC, but, may be substituted for by PMA. Both stimuli must occur simultaneously for at least 2–4 hr for the commitment of Jurkat to produce IL-2 (Weiss *et al.*, 1987).

Since two stimuli were required for IL-2 production by a single homogenous T-cell line, it was of interest to examine the cellular level upon which these two stimuli exerted their influence. IL-2 and IFN mRNA were undetectable in unstimulated Jurkat. Simultaneous stimulation with both T3 MAb and PMA were required for the accumulation of both IL-2 and IFN RNA and these genes appeared to be coordinately regulated (Weiss *et al.*, 1984b; Wiskocil *et al.*, 1985; Weiss *et al.*, 1987). These studies suggest that both stimuli exert their influence upon pretranslational events. More recent studies have suggested transcriptional regulation based on nuclear run-on experiments (Hardy *et al.*, 1986). In recent studies, the ability of cycloheximide to inhibit the appearance of IL-2 transcripts in stimulated Jurkat cells suggests that products of genes activated earlier may influence IL-2 transcription (Weiss *et al.*, 1987).

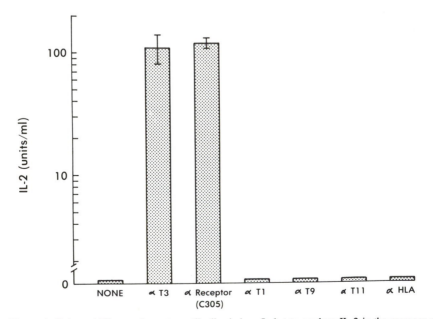

Figure 1. Only anti-T3 or antireceptor antibodies induce Jurkat to produce IL-2 in the presence of PMA. Cultures of Jurkat were prepared at 10^6 cells/ml with the indicated stimuli and 50 ng/ml PMA and after 24 hr cellfree supernatants were collected and assessed for IL-2 activity.

4. T3/Ti-Induced Hydrolysis of Polyphosphoinositides and Increase in the Concentration of Cytoplasmic Free Calcium

Recent attention has focused upon the mechanism of transmembrane signaling by cell-surface receptors. In the case of the T-cell antigen receptor, this is of particular interest because of the complexity of its structure, the additional requirements observed for T-cell antigen recognition and resultant activation, and the interest in T-cell regulatory phenomena (i.e., tolerance, suppression, autoimmunity, etc.). Studies of the response of the T3/Ti negative mutants of Jurkat (described above) to the plant lectin phytohemagglutinin (PHA) and concanavalin A (Con A) have helped to elucidate the mechanism of transmembrane signaling by the T3/Ti complex. When these mutants were stimulated with either PHA or Con A plus PMA, in contrast to the abundant levels of IL-2 produced by the parent cell line, minimal or no detectable IL-2 was produced (Weiss and Stobo, 1984; Weiss et al., 1984a; Weiss et al., 1987). This suggested that the expression of the T3/Ti complex was required for the activation of these cells by PHA or Con A. Alternatively, it was possible that these cells had lost the ability to produce IL-2 through independent mutational events. Therefore, it was necessary to demonstrate that these mutant cells retained their capacity to produce IL-2. Among the many biochemical events which have been associated with lymphocyte activation, increases in the concentration of cytoplasmic free calcium ($[Ca^{2+}]_i$) have been observed and can be reproduced with calcium ionophores. Calcium ionophores had been shown to be mitogenic for T lymphocytes under certain circumstances (Luckasen et al., 1974). Therefore, we examined the ability of calcium ionophores to bypass the requirement for perturbation of the T3/Ti complex in the activation Jurkat and the T3/Ti negative mutants. Indeed, Jurkat cells could be activated to produce abundant levels of IL-2 by either of two calcium ionophores, but only if they were added in the presence of PMA (Weiss and Stobo, 1984; Weiss et al., 1984a; Imboden et al., 1985a). The T3/Ti negative mutants of Jurkat could likewise produce abundant quantities of IL-2 if stimulated with a calcium ionophore plus PMA (Weiss and Stobo, 1984; Weiss et al., 1984a). Moreover, in a cell derived from a β-chain deficient mutant of Jurkat, in which T3/Ti expression was reconstituted by gene transfer (Ohashi et al., 1985), the response to PHA and Con A was restored (Weiss et al., 1987). These studies strongly argue for the expression of the T3/Ti complex as a requirement for the activation of Jurkat by PHA or Con A. More importantly, however, they suggested that the T3/Ti complex might initiate activation by increasing $[Ca^{2+}]_i$.

The development of calcium sensitive dyes, such as quin2, has permitted the measurement of $[Ca^{2+}]_i$ in lymphoid cells (Tsien et al., 1982). We utilized quin2 to monitor changes in $[Ca^{2+}]_i$ in Jurkat cells in response to various stimuli. When Jurkat cells were stimulated with anti-T3, anti-Ti, PHA, or calcium ionophores, prompt three- to five-fold increases $[Ca^{2+}]_i$ were observed (Fig. 2)

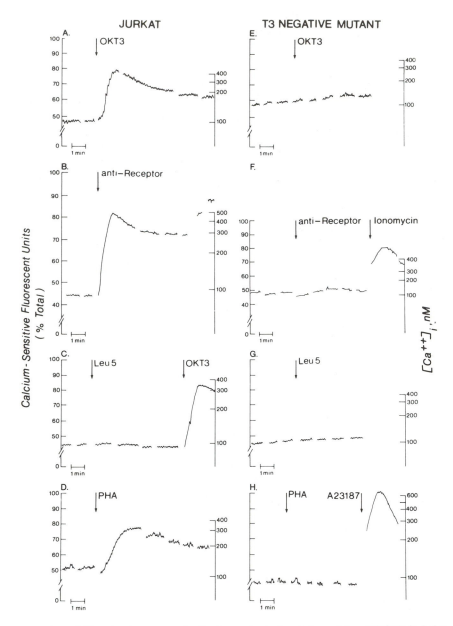

Figure 2. Anti-T3 and antireceptor antibodies increase cytoplasmic free calcium ([Ca²⁺]ᵢ) in Jurkat cells. Cells were loaded with quin2, and the fluorescence of the cellular suspension was measured over the indicated time period after addition of the indicated stimulus.

(Weiss *et al.*, 1984a; Imboden *et al.*, 1985a). These increases in $[Ca^{2+}]_i$ were sustained for greater than 60 min. Moreover, the dose of ionophore required to activate Jurkat to produce IL-2 in the presence of PMA induced a rise in $[Ca^{2+}]_i$ comparable to those seen with activating doses of anti-T3 or anti-Ti MAb. Increases in $[Ca^{2+}]_i$ were not observed in response to PMA or to a variety of MAb reactive with other cell-surface molecules expressed on Jurkat (i.e., MAb reactive with T1, T4, T9, or HLA-A,B,C) (Weiss *et al.*, 1984a; Imboden *et al.*, 1985a). These results have been confirmed in other laboratories utilizing other cell lines as well as normal resting peripheral blood T cells (O'Flynn *et al.*, 1985; Oettgen *et al.*, 1985). More recent studies have also demonstrated that calcium ionophores, in the presence of PMA, are mitogenic for peripheral human T cells (Hara and Fu, 1985). Collectively, these results strongly argue that the T3/Ti complex initiates activation through mechanisms which increase $[Ca^{2+}]_i$.

We also investigated the failure of the T3/Ti-negative mutants of Jurkat to respond to PHA or Con A by examining the responses of quin2-loaded mutant cells. As expected, no change in $[Ca^{2+}]_i$ was observed in quin2-loaded T3/Ti-negative mutant cells in response to anti-T3 or anti-Ti (Fig. 2). Likewise, PHA also failed to increase $[Ca^{2+}]_i$ in these cells, supporting the notion that activation of Jurkat with PHA requires the cell-surface expression of the T3/Ti complex (Fig. 2) (Weiss *et al.*, 1984a). Similar results have been obtained with Con A (Weiss *et al.*, 1987).

The mechanism by which the T3/Ti complex increases $[Ca^{2+}]_i$ has been studied. Increases in $[Ca^{2+}]_i$ may be a consequence of transmembrane flux of Ca^{2+} across the plasma membrane or release of Ca^{2+} from intracellular stores. It has been proposed that T3 may form a calcium channel and this could account for the observed increase in $[Ca^{2+}]_i$ following interaction of the T3/Ti complex with the appropriate ligand (Oettgen *et al.*, 1985). However, we demonstrated that the initial increase in $[Ca^{2+}]_i$ observed during the first two minutes in response to anti-Ti could occur when extracellular Ca^{2+} was chelated by EGTA to levels below those present intracellularly (Imboden and Stobo, 1985). This suggests that the initial increase in $[Ca^{2+}]_i$ can be accounted for by the release of internal stores. It is noteworthy that others have not observed this increase in $[Ca^{2+}]_i$ in the absence of extracellular Ca^{2+} (Oettgen *et al.*, 1985). However, this discrepancy may be due to the Ca^{2+} buffering capacity of quin2. When used at high concentrations, quin2 can buffer released Ca^{2+} and prevent T3/Ti-mediated increases in $[Ca^{2+}]_i$ due solely to released Ca^{2+} (Imboden *et al.*, 1985b). On the other hand, regardless of cellular quin2 concentrations, sustained T3/Ti-mediated increases in $[Ca^{2+}]_i$ require the presence of extracellular Ca^{2+} (Imboden and Stobo, 1985).

A variety of cellular receptors which increase $[Ca^{2+}]_i$ by the release of internal stores have been described and utilize a well-characterized signaling pathway involving the hydrolysis of phosphatidyl inositol bisphosphate (PIP_2). PIP_2, a relatively rare membrane phospholipid, is hydrolyzed by phospholipase C as a consequence of ligand–receptor interactions. This results in the formulation of two second messengers, inositol trisphosphate (IP_3) and diacylglycerol (DG). IP_3 can bind to specific receptors in the endoplasmic reticulum and induce the release of Ca^{2+} (Berride, 1984; Joseph *et al.*, 1984; Streb *et al.*, 1983; Thomas *et al.*, 1984). The effects of the other metabolite of PIP_2 hydrolysis, DG, will be discussed in detail below. An anti-Ti MAb was shown to induce a rapid increase in IP_3 in Jurkat (Imboden and Stobo, 1985). IP_3 added to permeabilized Jurkat cells resulted in the release of Ca^{2+} from internal stores (Imboden and Stobo, 1985). Moreover, antigen recognition by an antigen-specific, MHC-restricted human T-cell clone is associated with a substantial, sustained increase in IP_3 (Imboden *et al.*, 1987). Thus, the T3/Ti complex appears to participate in the initiation of T-cell activation by inducing the hydrolysis of PIP_2. The mechanism by which the perturbation of the T3/Ti complex induces PIP_2 hydrolysis is not clear. However, GTP-binding proteins which appear to regulate Ca^{2+}-mobilizing receptors in the neutrophil, may be involved in transmitting the signal from the T3/Ti complex to the phospholipase C. This is based on the observation that cholera toxin, which modifies GTP binding, can inhibit increases in IP_3 and $[Ca^{2+}]_i$ in Jurkat cells by a cAMP-independent mechanism (Imboden *et al.*, 1986). While it seems likely that the T3/Ti complex initiates the initial Ca^{2+} by inducing the hydrolysis of PIP_2, the details regarding the maintenance of increased levels of Ca^{2+} are not clear. Moreover, the relative function of T3, in contrast to the presumed antigen-binding function of the Ti heterodimer, remains to be elucidated.

5. T-Cell Activation Induced by Soluble or Immobilized T3 MAb

As discussed, under appropriate conditions MAb reactive with the T3/Ti complex can mimic the effects of antigen in the activation of T cells. Meuer *et al.* first demonstrated that the functional properties of T3 or Ti MAb depend upon whether they are present in soluble form of bound to a solid surface. They demonstrated that human T-cell clones could be activated to produce lymphokines and to proliferate in response to T3 or Ti MAb immobilized to sepharose beads (Meuer *et al.*, 1983b). Similarly, immobilized, but not soluble, anti-T3 is able to induce IL-2 production by the human T-cell line HUT 78 (Manger *et al.*, 1985). However, human T-cell clones as well as HUT 78 express activation markers such as HLA-DR antigens or the IL-2 receptor (TAC) (Manger *et al.*, 1985; Meuer *et al.*, 1984). Therefore, the ability to respond to immobilized anti-T3 alone may reflect the stimulation requirements of activated, rather than resting, T lymphocytes.

It is more difficult to investigate activation requirements for resting peripheral T cells because different numbers of contaminating AC in T-cell preparations obtained by different purification methods could account for variations observed in the response to various stimuli. Unfractionated human PBL proliferate in response to soluble T3 MAb (Van Wauwe and Goossens, 1981). As discussed, this response requires the participation of AC contained in the adherent cell population (Van Wauwe and Goossens, 1981; Kaneoka et al., 1983). One function of these AC in the T3-induced activation of T cells is to immobilize anti-T3 MAb via their Fc receptors (Tax et al., 1983). When the majority of AC are removed from PBL by classical adherence methods, they can no longer be activated by soluble anti-T3 (Van Wauwe et al., 1980). However, they still proliferate in response to immobilized T3 MAb (Table II). Moreover, as described, the phorbol ester, PMA, can also reconstitute the response of adherent cell-depleted PBL to soluble anti-T3 (Table I) (Hara and Fu, 1985). In contrast, when highly purified T cells are isolated by more stringent AC depletion methods, they do not proliferate in response to soluble or immobilized T3 MAb unless PMA is added (Table II) (Schwab et al., 1985; Palacios, 1985). The inability of such purified T cells to respond to immobilized anti-T3 suggests that AC cells might provide a second function in addition to and distinct from their immobilization of T3 MAb.

Since the second function provided by AC appears to be provided by small numbers of such cells remaining after adherence protocols, it seemed likely that this function involved the secretion of an AC-derived soluble factor. Interleukin-1 (IL-1) is a soluble product of macrophages that, under some conditions, synergizes in the activation of T cells (Mizel, 1982). Therefore, we examined the ability of IL-1 to reconstitute the response of highly purified T cells to immo-

Table II. Fractions of Resting T Cells Obtained by Different Purification Methods Differ in Their Stimulation Requirements[a]

Responding fraction	Stimulus	None	+IL-1	+9.3
Unfractionated PBL	None	1,029	2,131	571
	Soluble OKT3	72,900	93,312	79,336
	Immobilized OKT3	87,193	96,185	102,348
Nonadherent PBL	None	250	571	506
	Soluble OKT3	520	1,640	2,612
	Immobilized OKT3	77,388	96,185	126,453
Highly purified T cells	None	171	821	215
	Soluble OKT3	411	461	533
	Immobilized OKT3	2,974	24,346	102,954

[a]The indicated lymphocyte populations, were stimulated for 96 hrs with the following concentration of reagents: OKT3 ascites at a 1:1000, 9.3 at 1:10,000 dilution of ascites, and IL-1 at 100 U/ml. For stimulation with immobilized OKT3, culture wells were precoated with goat anti-mouse immunoglobulin at pH 9.5. The concentration of cells during culture was 1×10^6/ml for unfractionated cells and non-adherent PBL, 5×10^5/ml for highly purified T cells. Figures represent c.p.m. [^{3}H]-thymidine uptake (mean values of a representative experiment; standard deviation less than 15%).

bilized T3 MAb. Indeed, the response to such highly purified T cells to immobilized, but not soluble, anti-T3 was reconstituted if IL-1 was added (Table II). Similar observations have been made in other laboratories (Schwab *et al.*, 1985; Palacios, 1985; Williams *et al.*, 1985; Scheurich *et al.*, 1985). These findings suggest that AC have a dual role in the T3-induced activation of peripheral T cells. One function is the immobilization of anti-T3 via the Fc receptors on such AC. The second function may be the secretion of IL-1. Both of these functions can apparently be bypassed by PMA.

   A prediction from such findings would be that T-cell surface molecules other than the T3/Ti complex may function as receptors for such second signals provided by AC. Therefore, we screened a large panel of MAb reactive with T-cell surface molecules for their ability to synergize with immobilized anti-T3 MAb. One such MAb, 9.3, which reacts with most peripheral T cells and recognizes an 80- to 90-kD homodimer, was able to provide this agonist function (Table II) (Hansen *et al.*, 1980; Clark *et al.*, 1983). Thus, 9.3 could synergize with immobilized anti-T3 but not with soluble anti-T3 or a calcium ionophore (Ledbetter *et al.*, 1985; Weiss *et al.*, 1986). It would be tempting to speculate that 9.3 reacts with the IL-1 receptor, but the size of the putative IL-1 receptor (based on crosslinking studies) is distinctly different from Tp44, the cell surface molecule recognized by 9.3 (Dower *et al.*, 1985). However, based on the similarity in the biological effects of IL-1 and 9.3, it is likely that these two reagents may initiate similar, as yet unidentified, biochemical events via perturbation of their respective cell-surface receptors.

   The T-cell leukemia, Jurkat, which phenotypically resembles a resting T cell, also functionally mimics the highly purified T cells described above in its requirements for stimulation (Manger *et al.*, 1985). Thus, Jurkat cells secrete IL-2 in response to immobilized, but not soluble, anti-T3 if either IL-1 or 9.3 is added (Manger *et al.*, 1985; Weiss *et al.*, 1986). Therefore, Jurkat can serve as a model to investigate differences in signaling events in T cells in an AC-free system. Since PMA can substitute for immobilization of anti-T3 in T-cell activation and PMA is a potent activator of protein kinase C (pkC), we examined the activation of this enzyme in Jurkat cells after stimulation with soluble or immobilized anti-T3.

6. The Role of Protein Kinase C in the Activation of Resting T Lymphocytes

   A variety of hormones and growth factors induce the activation of phospholipase C in their target cells upon interaction with specific surface receptors (Mitchell, 1975; Prpic *et al.*, 1982). As described above, this signaling mechanism is also activated when the T3/Ti complex on T lymphocytes is triggered by MAbs against this structure. Phospholipase C catalyzes the hydrolysis PIP_2 to IP_3 and diacylglycerol (DG). As discussed in detail above, IP_3 mediates the initial increase in $[Ca^{2+}]_i$. In contrast, the other metabolite, DG, activates a Ca^{2+} and phospholipid-dependent kinase, termed pkC, by markedly increasing the affinity of this enzyme for Ca^{2+} and phospholipids (Kishimoto *et al.*, 1980; Majerus *et al.*, 1984). In unstimulated cells, pkC is present as an inactive form in the cytoplasm. The receptor-mediated generation of DG induces a transloca-tion and attachment of pkC to the cell membrane, whereupon contact with phos-pholipids and Ca^{2+} it becomes enzymatically active (Fig. 3A) (Kraft and Anderson, 1983). This, however, is a reversible process, depending on the amount of receptor-generated DG and its degradation rate. DG can be readily metabolized, either by further hydrolysis to monoacylglycerol and arachidonic acid or by phosphorylation to phosphatidic acid, which eventually gets recycled to phosphatidyl inositols (Nishizuka, 1984a). Indeed, it has previously been shown that stimulation of T cells with soluble anti-T3 causes a rapidly reversible translocation of pkC from the cytoplasm to the membrane (Farrar and Ruiscetti, 1986). Thus, the interaction of antibodies with the T3/Ti complex on T cells induces both a rise in $[Ca^{2+}]_i$ and a transient activation of pkC. It is widely accepted that $[Ca^{2+}]_i$ increase and pkC activation are synergistically acting in-tracellular signals, which induce functional activation in a variety of cells (i.e., platelets, mast cells, neutrophils, pancreatic cells, etc.) (Michell, 1983; Nishizuka, 1984b). Therefore, it is apparently a paradox that T3 antibodies by

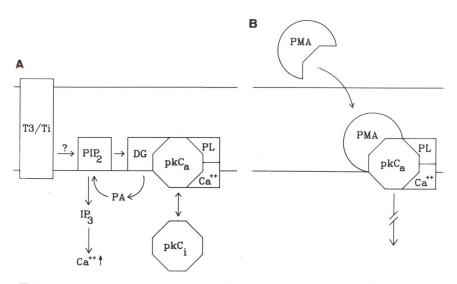

Figure 3. Model of T3/Ti-induced (A) and PMA-induced (B) activation of pkC. T3/Ti, T-cell receptor complex; PIP_2, phosphatidylinositol bisphosphate; IP_3, inositol trisphosphate; DG, di-acylglycerol; PA, phospatidic acid; PL, phospholipid; pkC_a, active form of protein kinase C; pkC_i, inactive form of protein kinase C; PMA, phorbol-myristate-acetate.

themselves are not sufficient to induce activation of T cells, as measured by proliferation or lymphokine production. The T-cell line Jurkat requires, in addition to T3 antibodies, the presence of PMA as a second stimulus (Weiss *et al.*, 1984a,b,c). for the induction of IL-2 or IFN-γ secretion. The highly lipophilic PMA is able to intercalate into the membrane and to mimic effects of DG by mediating a high affinity interaction between pkC and membrane (Fig. 3B). Treatment of Jurkat cells with PMA results in a loss of pkC activity from the cytosol and a significant increase of pkC activity associated with the plasma membrane within minutes. Since PMA, in contrast to DG, cannot be readily metabolized, the PMA-induced pkC translocation is virtually irreversible (Kraft and Anderson, 1983; Farrar and Anderson, 1985). The requirement for PMA in addition to anti-T3 for the activation of Jurkat suggests that the transient anti-T3-induced activation of pkC is not sufficient to result in lymphokine production. However, when Jurkat cells were stimulated with immobilized rather than soluble T3 antibodies, the duration of pkC activation was prolonged (Table III). After stimulation with soluble T3, pkC translocation was completely reversed within 10 min. In cells stimulated with immobilized anti-T3, a substantial proportion of pkC activity remained translocated to the membrane for up to 2 hr (Manger *et al.*, 1987). This prolonged pkC activation might be necessary for phosphorylation of increased amounts of specific substrate(s) required for further intracellular events in T-cell activation.

Table III. Stimulation of Jurkat Cells with Immobilized OKT3 Induces a Prolonged Translocation of Protein Kinase C[a]

Stimulus	Duration	Protein kinase C activity (ratio of cytosol/membrane)
Unstimulated		14.6
PMA	10 min	0.4
Soluble OKT3	1 min	2.5
	3 min	4.7
	10 min	13.8
	2 hr	11.7
Immobilized OKT3	3 min	3.0
	10 min	5.0
	2 hr	4.8

[a]Stimulation of 10^7 Jurkat cells was performed in plastic culture dishes that had been coated with goat anti-mouse immunoglobulin for immobilization of OKT3 or untreated dishes for soluble OKT3. Concentration for OKT3 antibodies is 1:1000 ascites dilution, and 50 ng/ml for PMA. The ratio of cytosolic to membrane pkC activity is shown for one representative experiment, which was performed in duplicates. The enzyme activity was determined as described (Kraft *et al.*, 1982).

To investigate further the role of pkC activation in T-cell activation, we used a synthetic DG, dioctanoylglycerol (DiC8) because a recent report suggested that PMA has important intracellular effects other than the activation of pkC (Yamamoto et al., 1985). DiC8 is cell-permeable and induces translocation and activation of pkC. However, in contrast to the irreversible pkC activation by PMA, the DiC8-induced effect starts to disappear after about 30 min, depending on the concentration of DiC8 added (May et al., 1986). This is presumably due to the fact that, once DiC8 has entered the cell, it is phosphorylated much like endogenous DG. The metabolite dioctanoyl phosphatidic acid can be detected in a monocytic cell line 10 min after stimulation with DiC8 (May et al., 1985). Therefore, the duration and magnitude of the DiC8-induced pkC activation resembles that induced by endogenous DG generated after stimulation of Jurkat cells with immobilized anti-T3. The combination of $[Ca^{2+}]_i$ increase by ionomycin and DiC8-induced pkC activation is sufficient to stimulate IL-2 receptor expression and proliferation of adherent cell depleted T lymphocytes, which contain approximately 1% residual AC (Manger et al., 1987). These results provide further support for an important role of pkC in T-cell activation. Thus, three factors may influence activation of resting T cells by T3/Ti MAb: (1) Triggering of the T3/Ti complex; (2) surface events influencing redistribution of T3/Ti (i.e., by immobilization of stimulating antibodies), which leads to PIP_2 hydrolysis and to a prolonged translocation of pkC; and (3) IL-1, or the ligand of Tp44, inducing an unknown intracellular signal, which can also be provided by PMA.

7. Summary

Activation of T lymphocytes during antigen recognition requires the interaction of the T-cell receptor complex (T3/Ti) with antigen presented on the surface of accessory cells (AC) in the context of self-MHC. Under certain conditions monoclonal antibodies reactive with T3/Ti can also induce T-cell activation. Soluble T3/Ti antibodies have been shown to induce the hydrolysis of polyphosphoinositides, which leads to the generation of inositol trisphosphate (IP_3) and diacylglycerol (DG). IP_3 triggers an initial increase in $[Ca^{2+}]_i$ by releasing Ca^{2+} from intracellular stores. The generation of DG, in contrast, leads to the activation of protein kinase C (pkC) by translocating it from the cytoplasm to the cell membrane. A rapid but transient increase in pkC activity seen in T cells stimulated with soluble T3 antibodies was markedly prolonged when these antibodies were immobilized to a solid surface. However, immobilized T3 antibodies are not sufficient to activate highly-purified resting T cells. These cells require, in addition, the presence of interleukin-1 for the induction of proliferation. An IL-1-like functional effect can be demonstrated for the monoclonal antibody 9.3, which reacts with an 80- to 90-kd homodimer on the majority of T cells. The natural ligand for this surface molecule remains to be identified.

ACKNOWLEDGMENTS. The authors wish to thank Dr. John Stobo for his support and advice during the course of these studies and Michael Armanini for his excellent secretarial assistance.

References

Allison, J.P., and Lanier, L.L., 1985, Identification of antigen receptor-associated structures on murine T cells, *Nature* **314:**107.

Allison, J., McIntyre, B., and Bloch, D., 1982, Tumor-specific antigen of murine T-lymphoma defined with monoclonal antibody, *J. Immunol.* **129:**2293.

Berride, M.J., 1984, Inositol trisphosphate and diacylglycerol as second messengers, *Biochem. J.* **220:**345.

Borst, J., Alexander, S., Elder, J., and Terhorst, C., 1983, The T3 complex on human T lymphocytes involves four structurally distinct glycoproteins, *J. Biol. Chem.* **258:**5135.

Brenner, M.B., Trowbridge, I.S., and Strominger, J.L., 1985, Cross-linking of human T cell receptor proteins: Association between the T cell idiotype β subunit and the T3 glycoprotein heavy subunit, *Cell* **40:**183.

Clark, E.A., Martin, P.J., Hansen, J.A., and Ledbetter, J.A., 1983, Evolution of epitopes on human and non human primate lymphocyte cell surface antigens, *Immunogenet.* **18:**599.

Dower, S.K., Kronheim, S.R., March, C.J., Conlon, P.J., Hopp, T.P., Gillis, S., and Urdal, D.L., 1985, Detection and characterization of high affinity plasma receptors for human interleukin 1, *J. Exp. Med.* **162:**501.

Farrar, W.L., and Anderson, W.B., 1985, Interleukin-2 stimulates association of protein kinase C with plasma membrane, *Nature* **315:**233.

Farrar, W.L., and Ruiscetti, F.W., 1986, Association of protein kinase C activation with IL-2 receptor expression, *J. Immunol.* **136:**1266.

Farrar, J., Mizel, S., Fuller-Farrar, J., Farrar, W., and Hilfiker, M., 1980, Macrophage-independent activation of helper T cells, *J. Immunol.* **125:**793.

Gold, D.P., Puck, J.M., Pettey, C.L., Cho, M., Coligan, J., Woody, J.N., and Terhorst, C., 1986, Isolation of cDNA clones encoding the 20K non-glycosylated polypeptide chain of the human T-cell receptor/T3 complex, *Nature* **321:**431.

Hansen, J.A., Martin, P.J., and Nowinski, C., 1980, Monoclonal antibodies identifying a novel T-cell antigen and Ia antigens of human lymphocytes, *Immunogenet.* **10:**247.

Hara, T., and Fu, S.M., 1985, Human T cell activation. I. Monocyte-independent activation and proliferation induced by anti-T3 monoclonal antibodies in the presence of tumor promoter 12-0-tetradecanoyl phorbol-13-acetate, *J. Exp. Med.* **161:**641.

Hardy, K.J., Manger, B., Newton, M., and Stobo, J.D., 1987, Molecular events involved in regulating human interferon-γ gene expression during T cell activation, *J. Immunol.* **138:**2353.

Haskins, K., Kubo, R., White, J., Pigeon, M., Kappler, J., and Marrack, P., 1983, The major histocompatibility complex-restricted antigen receptor on T cells, *J. Exp. Med.* **157:**1149.

Imboden, J.B., and Stobo, J.D., 1985, Transmembrane signaling by the T cell antigen receptor, *J. Exp. Med.* **161:**446.

Imboden, J., Weiss, A., and Stobo, J.D., 1985a, The antigen receptor on a human T cell line initiates activation by increasing cytoplasmic free calcium, *J. Immunol.* **134:**663.

Imboden, J.B., Weiss, A., and Stobo, J.D., 1985b, Transmembrane signaling by the T3-antigen receptor complex, *Immunol. Today* **6:**328.

Imboden, J.B., Shoback, D.M., Pattison, G., and Stobo, J.D., 1986, Cholera toxin inhibits the T cell antigen receptor-mediated increases in inositol trisphosphate and cytoplasmic free calcium, *Proc. Natl. Acad. Sci. USA* (in press).

Imboden, J., Weyand, C., and Goronzy, J., 1987, Antigen recognition by a human T cell clone leads to increases in inositol trisphosphate, *J. Immunol.* **138**:1322.

Joseph, S.K., Thomas, A.P., Williams, R.J., Irvine, R.F., and Williamson, J.R., 1984, Myo-inositol-1,4,5-trisphosphate. A second messenger for the hormonal mobilization of intracellular Ca^{2+} in liver, *J. Biol. Chem.* **259**:3077.

Kaneoka, H., Perez-Rojas, G., Sasasuki, T., Benike, C.J., and Engelman, E.G., 1983, Human T lymphocyte proliferation induced by a pan-T monoclonal antibody (anti-Leu 4): heterogeneity of response is a function of monocytes, *J. Immunol.* **131**:158.

Kaye, J., Porcelli, S., Tite, J., Jones, B., and Janeway, Jr., C., 1983, Both a monoclonal antibody and antisera specific for determinants unique to individual cloned helper T cell lines can substitute for antigen and antigen-presenting cells in the activation of T cells, *J. Exp. Med.* **158**:836.

Kishomoto, A., Takai, Y., Mori, T., Kikkawa, U., and Nishizuka, Y., 1980, Activation of calcium and phospholipid-dependent protein kinase by diacylglycerol, its possible relation to phosphatidylinositol turnover, *J. Biol. Chem.* **255**:2273.

Kraft, A.S., and Anderson, W.B., 1983, Phorbol esters increase the amount of Ca^{2+}, phospholipid-dependent protein kinase associated with plasma membrane, *Nature (Lond.)* **301**:621.

Kraft, A.S., Anderson, W.B., Cooper, H.L., and Sando, J.J., 1982, Decrease in cytosolic calcium/phospholipid-dependent protein kinase activity following phorbol ester treatment of EL4 thymoma cells, *J. Biol. Chem.* **257**:13193.

Krissansen, G.W., Owen, M.J., Verbi, W., and Crumpton, M.J., 1986, Primary structure of the T3γ subunit of the T3/T cell antigen receptor complex deduced from cDNA sequences: evolution of the T3γ and δ subunits, *EMBO J.* **5**:1799.

Ledbetter, J.A., Martin, P.J., Spooner, C.E., Wofsy, D., Tsu, T.T., Beatty, P.G., and Gladstone, P., 1985, Antibodies to Tp67 and Tp44 augment and sustain proliferative response of activated T cells, *J. Immunol.* **135**:2331.

Luckasen, J.R., White, J.G., and Kersey, J.H., 1974, Mitogenic properties of a calcium ionophore, A23187, *Proc. Natl. Acad. Sci. USA* **71**:5088.

Majerus, P.W., Neufeld, E.J., and Wilson, D.B., 1984, Production of phosphoinositide-derived messengers, *Cell* **37**:701.

Manger, B., Weiss, A., Weyland, C., Goronzy, J., and Stobo, J.D., 1985, T cell activation: differences in the signals required for IL 2 production by nonactivated and activated T cells, *J. Immunol.* **135**:3669.

Manger, B., Weiss, A., Imboden, J., Laing, T., and Stobo, J., 1987, The role of protein kinase C in transmembrane signaling by the T cell antigen receptor complex. Effects of stimulation with soluble or immobilized CD3 antibodies, *J. Immunol.* **139**:2755.

May, W.S., Lapetina, E.G., and Cuatrecasas, P., 1986, Intracellular activation of protein kinase C and regulation of the surface transferrin receptor by diacylglycerol is a spontaneously reversible process that is associated with rapid formation of phosphatidic acid, *Proc. Natl. Acad. Sci. USA* **83**:1281.

Meuer, S., Fitzgerald, K., Hussey, R., Hodgdon, J., Schlossman, S., and Reinherz, E., 1983a, Clonotypic structures involved in antigen-specific human T cell function, *J. Exp. Med.* **157**:705.

Meuer, S., Hodgdon, J., Hussey, R., Protentis, J., Schlossman, S., and Reinherz, E., 1983b, Antigen-like effects of monoclonal antibodies directed at receptors on human T cell clones, *J. Exp. Med.* **158**:988.

Meuer, S.C., Hussey, R.E., Cantrell, D.A., Hodgdon, J.C., Schlossman, S.F., Smith, K.A., and Reinherz, E.L., 1984, Triggering of the T3-Ti antigen receptor complex results in clonal T-cell proliferation through an interleukin-2 dependent autocrine pathway, *Proc. Natl. Acad. Sci. USA* **81**:1509.

Michell, B., 1983, Ca^{2+} and protein kinase C: two synergistic cellular signals, *Trends Biochem. Sci.* **8**:263.

Mitchell, R.H., 1975, Inositol phospholipids and cell surface receptor function, *Biochem. Biophys.*

    Acta. **415**:81.
Mizel, S.B., 1982, Interleukin 1 and T cell activation, *Immunol. Rev.* **63**:51.
Nishizuka, Y., 1984a, Turnover of inositol phospholipids and signal transduction, *Science* **225**:1365.
Nishizuka, Y., 1984b, The role of protein kinase C in cell surface signal transduction and tumor promotion, *Nature (Lond.)* **308**:693.
O'Flynn, K., Zanders, E., Lamb, J., Beverley, P., Wallace, D., Tatham, P., Tax, W., and Linch, D., 1985, Investigation of early T cell activation: Analysis of the effect of specific antigen, interleukin 2 and monoclonal antibodies on intracellular free calcium concentration, *Eur. J. Immunol.* **15**:7.
Oettgen, J., Terhorst, C., Cantley, L., and Rosoff, P., 1985, Stimulation of the T3-T cell receptor complex induces a membrane-potential-sensitive calcium influx, *Cell* **40**:583.
Ohashi, P.S., Mak, T.W., Van den Elsen, P., Yanagi, Y., Yoshikai, Y., Calman, A., Terhorst, C., Stobo, J.D., and Weiss, A., 1985, Reconstitution of an active surface T3/T-cell antigen receptor by DNA transfer, *Nature* **316**:606.
Palacios, R., 1985, Mechanisms by which accessory cells contribute in growth of resting T lymphocytes by OKT3 antibody, *Eur. J. Immunol.* **15**:645.
Prpic, V., Blackmore, P.F., and Exton, J.H., 1982, Phosphatidylinositol breakdown induced by vasopressin and epinephrine in hepatocytes is calcium dependent, *J. Biol. Chem.* **257**:11323.
Reinherz, E., Meuer, S., Fitzgerald, K., Hussey, R., Hodgdon, J., Acuto, O., and Schlossman, S., 1983, Comparison of T3-associated 49- and 42-kilodalton cell surface molecules on individual human T-cell clones: Evidence for peptide variability in T-cell receptor structures, *Proc. Natl. Acad. Sci. USA* **80**:4104.
Rosenstreich, D., and Mizel, S., 1979, Signal requirements for T lymphocyte activation. I. Replacement of macrophage function with phorbol myristate acetate, *J. Immunol.* **123**:1749.
Samelson, L., Germain, R., and Schwartz, R., 1983, Monoclonal antibodies against the antigen receptor on a cloned T-cell hybrid, *Proc. Natl. Acad. Sci. USA* **80**:6972.
Samelson, L.E., Harford, J.B., and Klausner, R.D., 1985, Identification of the components of the murine T cell antigen receptor complex, *Cell* **43**:223.
Samelson, L.E., Patel, M.D., Weissman, A.M., Harford, J.B., and Klausner, R.D., 1986, Antigen activation of murine T cells induces tyrosine phosphorylation of a polypeptide associated with the T cell antigen receptor, *Cell* **46**:1083.
Scheurich, P., Ucer, U., Wrann, M., and Pfizenmaier, K., 1985, Early events during primary activation of T cells: antigen receptor cross-linking and interleukin 1 initiate proliferative response of human T cells, *Eur. J. Immunol.* **15**:1091.
Schwab, R., Crow, M.K., Russo, C., and Weksler, M.E., 1985, Requirements for T cell activation by OKT3 monoclonal antibody: Role of modulation of T3 molecules and interleukin-1, *J. Immunol.* **135**:1714.
Streb, H., Irvine, R.F., Berride, M.J., and Schulz, J., 1983, Release of Ca^{2+} from a nonmitochondrial intracellular store in pancreatic acinar cells by inositol-1,4,5-tris-phosphate, *Nature* **306**:67.
Tax, W.J.M., Willems, H.W., Reekers, P.P.M., Capel, P.J.A., and Koene, R.A.P., 1983, Polymorphisms in mitogenic effect of IgG1 monoclonal antibodies against T3 antigen on human T cells, *Nature (Lond.)* **304**:445.
Thomas, A.P., Alexander, J., and Williamson, J.R., 1984, Relationship between inositol polyphosphate production and the increase of cytosolic free Ca^{2+} induced by vasopressin in isolated hepatocytes, *J. Biol. Chem.* **259**:5574.
Tsien, R.Y., Pozzan, T., and Rink, T.J., 1982, T-cell mitogens cause early changes in cytoplasmic free Ca^{2+} and membrane potential in lymphocytes, *Nature* **295**:68.
Van den Elsen, P., Shepley, B.-A., Borst, J., Coligan, J.E., Markham, A.F., Orkin, S., and Terhorst, C., 1984, Isolation of cDNA clones encoding the 20K T3 glycoprotein of human T-cell receptor complex, *Nature* **312**:413.

Van Wauwe, J.P., DeMey, J.R., and Goossens, J.G., 1980, OKT3: a monoclonal anti-human T lymphocyte antibody with potent mitogenic properties, *J. Immunol.* **124:**2708.

Van Wauwe, J.P., and Goossens, J.G., 1981, Mitogenic actions of orthoclone OKT3 on human peripheral blood lymphocytes: effects of monocytes and serum components, *Int. J. Immunopharmacol.* **3:**203.

Weiss, A., and Stobo, J.D., 1984, Requirement for the coexpression of T3 and the T cell antigen receptor on a malignant human T cell line, *J. Exp. Med.* **160:**1284.

Weiss, A., Imboden, J., Shoback, D., and Stobo, J., 1984a, Role of T3 surface molecules in human T cell activation: T3 dependent activation results in a rise in cytoplasmic free calcium, *Proc. Natl. Acad. Sci. USA* **81:**4169.

Weiss, A., Imboden, J., Wiskocil, R., and Stobo, J., 1984b, The role of T3 in the activation of human T cells, *J. Clin. Immunol.* **4:**165.

Weiss, A., Wiskocil, R.L., and Stobo, J.D., 1984c, The role of T3 surface molecules in the activation of human T cells: a stimulus requirement for IL 2 production reflects events occurring at a pretranslational level, *J. Immunol.* **133:**123.

Weiss, A., Manger, B., and Imboden, J., 1986, Synergy between the T3/antigen receptor complex and Tp44 in the activation of human T cells, *J. Immunol.* **137:**819.

Weiss, A., Shields, R., Newton, M., Manger, B., and Imboden, J., 1987, Ligand-receptor interactions required for commitment to the activation of the interleukin 2 gene, *J. Immunol.* **138:**2169.

Williams, J.M., Deloria, D., Hansen, J.A., Dinarello, C.A., Loertscher, R., Shapiro, H.M., and Strom, T.B., 1985, The events of primary T cell activation can be staged by use of Sepharose-bound anti-T3 (64.1) monoclonal antibody and purified IL-1, *J. Immunol.* **135:**2249.

Wiskocil, R., Weiss, A., Imboden, J., Kamin-Lewis, R., and Stobo, J., 1985, Activation of a human T cell line: a two-stimulus requirement in the pretranslational events involved in the coordinate expression of interleukin 2 and γ-interferon genes, *J. Immunol.* **134:**1599.

Yamamoto, S., Gotoh, H., Aizu, E., and Kato, R., 1985, Failure of 1-oleoyl-2-acetylglycerol to mimic the cell-differentiating action of 12-0-tetradecanoylphorbol 13-acetate in HL-60 cells, *J. Biol. Chem.* **260:**14230.

Yanagi, Y., Yoshikai, Y., Leggett, K., Clark, S.P., Aleksander, I., and Mak, T.W., 1984, A human T cell-specific cDNA clone encodes a protein having extensive homology to immunoglobulin chains, *Nature* **308:**145.

Yanagi, Y., Chan, A., Chin, B., Minden, M., and Mak, T.W., 1985, Analysis of cDNA clones specific for human T cells and the α and β chains of the T-cell receptor heterodimer from a human T-cell line, *Proc. Natl. Acad. Sci. USA* **82:**3430.

8

The Structure and Expression of the T-Cell α-, β-, and γ-Chain Genes in Human Malignancies

J. SLINGERLAND, H. GRIESSER, T.W. MAK, and M.D. MINDEN

1. Introduction

An essential feature of the malignant state is the clonal expansion of a single cell (Fialkow, 1976). This discovery has had a great impact upon our understanding of the development and progression of tumors. The presence of a clonal population of cells often has been ascertained by the detection of acquired karyotypic abnormalities in the malignant cells (Rowley, 1978), or the altered pattern of expression of the enzyme G-6-PD in appropriate heterozygous females (Fialkow, 1976). The introduction of the techniques of molecular biology has provided the investigator with a number of other means for detecting clonal populations of malignant cells (Korsmeyer *et al.*, 1983); this has been amply demonstrated for malignancies involving lymphocytes (Korsmeyer *et al.*, 1983; Arnold *et al.*, 1983; Sklar *et al.*, 1984). In this review we will describe the genes and the probes that have been used to determine the clonal nature of malignancies involving T lymphocytes and indicate, where possible, how these studies have altered our understanding of these diseases.

1.1. The T Lymphocyte

The T lymphocyte plays a central and predominant role in the specific immune response. In order to carry out its functions the T cell must be able to

J. SLINGERLAND, H. GRIESSER, T.W. MAK, and M.D. MINDEN • Departments of Medicine and Medical Biophysics, Ontario Cancer Institute, University of Toronto, Toronto, Ontario M4X 1K9, Canada.

recognize a wide variety of antigens. As described elsewhere in this monograph antigens are recognized by the surface T-cell antigen receptor complex consisting of the T3 molecule (Borst *et al.*, 1983; Weiss and Stobo, 1983; Ohashi *et al.*, 1985) and the dimeric T-cell antigen receptor (TcR) (Acuto and Reinherz, 1985; Kapler *et al.*, 1983; Kronenberg *et al.*, 1986). The TcR is made up of two dissimilar proteins, the α chain and the smaller, more basic β chain. The generation of the required diversity of the receptor is accomplished through somatic recombination, in which one part of the T-cell receptor gene is physically dissociated from its normal position in the genome (germline configuration) and reassociated with another part of the T-cell antigen receptor gene (Yanagi *et al.*, 1984; Hedrick *et al.*, 1984a; Hedrick *et al.*, 1984b; Kronenberg *et al.*, 1986).

These rearrangements of the T-cell antigen receptor genes are a necessary and a normal part of the development of the T cell. Rearrangements occur in an apparently random manner (Yanagi *et al.*, 1984; Toyanaga *et al.*, 1984; Rabbitts *et al.*, 1985a; Kronenberg *et al.*, 1986) thus any given developing T cell will possess a unique rearrangement that can be considered to be its signature or name. Once rearranged, the structure of the gene appears to be fixed so that any progeny of a cell that has rearranged the α- or β-chain genes will carry the same rearrangement as the parent cell. One can then consider the rearrangement to represent the family name.

Rearrangements of the TcR genes are detected by the Southern blot technique (Southern, 1975). This is a relatively insensitive technique in that in order to detect a band on an autoradiograph there must be at least 10^4 copies of the same rearranged gene present in the sample (Minden *et al.*, 1985). In a typical experiment, DNA from 10^6 cells or 2×10^6 gene copies is loaded on the gel. In situations in which all of the cells have the same rearrangement the new pattern is readily detected, while in cases in which each cell bears a different rearrangement no bands are detected as the representation of any one particular pattern is below the level of detection. However, if one of the cells in this mixed population proliferates so that it now represents approximately 1% of the population of cells, then the rearranged gene of this "family" of cells can be detected. Figure 1 illustrates the qualitative and quantitative nature of the Southern blot in detecting rearrangements of the TcR_β gene.

2. T-Cell Malignancies

T-lymphocyte malignancies may take a number of different forms. They may resemble leukemias, involving mainly the bone marrow and the peripheral blood or they may present as lymphomas, involving mainly the lymph nodes. Until recently the clonal nature of an expanded population of T lymphocytes could only be resolved by karyotypic analysis. The distinction between a clonal

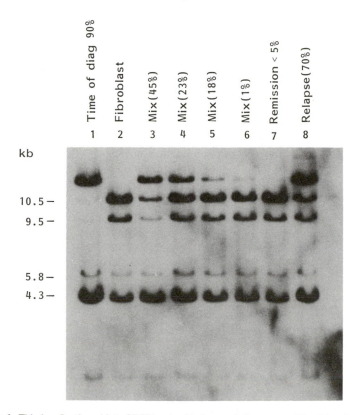

Figure 1. This is a Southern blot of DNA cut with the restriction enzyme Eco R1 and probed with ^{32}P-labeled DNA encoding the constant region of the β-chain gene. The DNA was derived either from the leukemic cells or normal fibroblasts of the same individual. These samples are in two leftmost lanes. The middle lanes represent mixtures of the normal and leukemic DNA in various proportions; the percent leukemic DNA is at the top of the lane. The lane labeled less than 5% blasts was a sample obtained at the time of remission while the last lane on the right was obtained at a time of marrow relapse.

population of T cells as compared to a polyclonal population of T cells reacting to an antigen is important in understanding disease process and possibly in making therapeutic decisions. With the isolation of the TcR genes it is now possible to use Southern blots to determine the clonality of expanded populations of T cells (Yanagi *et al.*, 1984; Hedrick *et al.*, 1984a; Yanagi *et al.*, 1985; Yoshikai *et al.*, 1985).

3. Technical Considerations

The ability to determine the clonal nature of a population of T cells is the result of somatic rearrangements that occur in all T cells. Three different genes have been found to undergo somatic rearrangement during the development of the mature T cell; they are the α- (Yoshikai *et al.*, 1985), β- (Yanagi *et al.*, 1984), and γ- (Lefranc *et al.*, 1986a,b; Lefranc and Rabbitts, 1985; Quertermous *et al.*, 1986) chain genes. The α and β chains make up part of the T-cell antigen receptor. The function of the γ chain is not known, however it has been demonstrated that the γ chain gene encodes for a protein of 35,000 MW and is associated with a second protein of 45,000 MW (Bank *et al.*, 1986; Brenner *et al.*, 1986). These two proteins, like the α- and β-chain proteins, are expressed on the surface of the T cell in association with the T3 antigen complex.

The genomic organization of the α-, β-, and γ-chain genes is similar to each other and to that of the immunoglobulin genes; these genes being composed of discontinuous sequences that encode for variable (V), joining (J), and constant (C) regions (Tonegawa, 1983; Kronenberg *et al.*, 1986; Hedrick *et al.*, 1984b; Sim *et al.*, 1984). The β chain also contains diversity (D) regions (Clark *et al.*, 1984). During the development of the T cell the V, D, and J regions are brought together through somatic recombination. The genomic structure of these three genes is depicted in Fig. 2.

The β chain is located on chromosome 7 band q34 (Caccia *et al.*, 1984; Morton *et al.*, 1985; Isobe *et al.*, 1985). It consists of approximately 50 V regions (Kimura *et al.*, 1986) and two C regions, $C_\beta 1$ and $C_\beta 2$ (Toyanaga *et al.*, 1985; Sims *et al.*, 1984). Located approximately 5 kb 5' of each of the constant regions is a cluster of 6 or 7 J regions. Approximately 650 bp 5' of each J region is at least one D region.

The α locus is located on the chromosome 14 bands q11–13 (Caccia *et al.*, 1985; Collins *et al.*, 1985; Croce *et al.*, 1985). There are approximately 50 V regions and one C region (Yoshikai *et al.*, 1986). Unlike the β chain there are more than 20 different J regions spread out over more than 80 kb 5' of the constant region (Yoshikai *et al.*, 1985). It is not known whether the α chain contains any D region segments.

The γ locus is located on chromosome 7 band p13 (Murre *et al.*, 1985). There are approximately 9 V regions (LeFranc *et al.*, 1986b) and 2 constant-region genes in man (Lefranc and Rabbits, 1985). 5' of each of the constant regions are the J segments.

During the process of rearrangement of the β chain, a D segment recombines with a J segment, the genetic material between these two sites is excised and usually lost from the genome. A V segment may then recombine with the rearranged D segment giving rise to a complete VDJ piece. Again, the intervening length of DNA is usually lost from the genome. The resulting gene may now be capable of producing a functional β-chain protein (Siu *et al.*, 1984). Similar

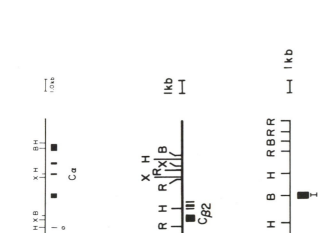

Figure 2. The structure of the (a), α-, (b), β-, and (c), γ-, chain genes are illustrated.

rearrangements occur for the α- and γ-chain genes, except it is not known whether there are D regions in these two genes. As a result of the rearrangement the germline structure is disrupted and is replaced by a new structure in which the location of the restriction enzyme sites have been altered. By using specific restriction enzymes and defined probes it is possible to detect such rearrangements. The probes and restriction enzymes used to study the structure of these three genes in T-cell proliferations are described.

3.1. Analysis of the α Chain

Studies of rearrangements of α-chain genes have been limited due to the great distance over which the J segments are dispersed. Using standard agarose electrophoretic techniques and commonly used restriction enzymes, probes to the constant region of the α chain only rarely detect rearrangements. However it has been possible to detect approximately 70% of expected rearrangements of this region using a series of probes derived from the 35 kb immediately 5' of the constant region (Sangster *et al.*, 1986; Baer *et al.*, 1986). In order to detect all rearrangements it will be necessary to isolate the entire J region and to generate representative single copy probes. A second way to detect rearrangements of the α region will be to use other forms of electrophoresis and restriction enzymes that cut infrequently. Recently, electrophoretic techniques that make it possible to separate very large fragments of DNA on the order of several hundred kb have been described (Schwartz and Cantor, 1984). With these methods, it should be possible to use a few well-characterized probes to detect α-chain rearrangements.

Transcription of the α-chain gene may result in either a 1.6- or 1.3-kb transcript (Yanagi *et al.*, 1985). The 1.6-kb transcript contains VJ and C segments and results from transcription of a rearranged gene. The 1.3-kb transcript contains J and C segments and is likely transcribed from the nonrearranged gene. Both transcripts have been found in leukemia T-cell lines and functional T-cell lines. Several authors have used the presence of a 1.6-kb message as evidence of gene rearrangement (Furley *et al.*, 1986). This assumption may be incorrect since: (1) 1.6-kb α mRNA has been found in cells that one would not expect to have rearranged their α-chain genes; and (2) in cells such as the leukemic cell line Molt 4 that are known to have rearranged their α-chain genes, no α mRNA was detected (Sangster *et al.*, 1986).

3.2. Analysis of the β Chain

In contrast to the α-locus, rearrangements of the β chain are detected readily. The coding portions of the constant region are highly homologous, while the segments 5' of the constant region are divergent, thus by using an appropriate combination of probe and restriction enzyme it is possible to determine whether the arrangement involves either the $C_\beta 1$ or the $C_\beta 2$ segment (Toyanaga *et al.*, 1985).

The $J_\beta 1$, $C_\beta 1$, $J_\beta 2$, and $C_\beta 2$ regions are contained on a 23-kb Bam H1 fragment in germline DNA. Rearrangements involving either the $\beta 1$ or $\beta 2$ regions will result in the loss of the 23-kb band and the appearance of a new band. This makes the combination of Bam H1 and a constant region probe very useful. However, due to the large size of the fragment and the fact that DNA is separated on an agarose gel in a logarithmic manner, small changes in the size of a fragment may be missed. This may be overcome by running low percentage gels for long periods of time at low voltage to obtain maximum resolution, or by using newer forms of electrophoresis, such as pulse field electrophoresis, which tends to separate large fragments in a linear, rather than a logarithmic manner, or by using another combination of probe and restriction enzyme that results in smaller fragments. The latter approach not only overcomes the problem of the large size of the DNA fragment, but also provides information as to whether the rearrangement involves $C_\beta 1$ or $C_\beta 2$.

DNA cleaved with Eco R1 and probed with a constant-region probe results in two bands; an 11-kb band that contains $C_\beta 1$ and a 4-kb band that contains $C_\beta 2$. Rearrangements involving $C_\beta 1$ results in the loss of the 11-kb band and the appearance of a new band. When $C_\beta 2$ is rearranged the $C_\beta 1$ region is usually lost from the genome, and thus the 11-kb Eco R1 band that contains $C_\beta 1$ is not seen on a Southern blot. Following such rearrangements the 4-kb Eco R1 band that contains $C_\beta 2$ persists and is not altered in size since there is an Eco R1 site between the $J_\beta 2$ and $C_\beta 2$.

Rearrangements of $C_\beta 2$ in DNA cleaved with Eco R1 may be detected by using a probe to the $J_\beta 2$ region. In the germline configuration a 4-kb band is observed. Rearrangement involving this region results in the loss of the 4-kb band and the appearance of a new band.

Rearrangement of $C_\beta 2$ may also be observed by probing Hind III cut DNA with a constant-region probe. This results in three germline bands of 8.0, 6.5, and 3.5 kb in length. The 3.5-kb band contains $C_\beta 1$ sequences, while the 8.0-kb band contains $J_\beta 2$ sequences and the 5' region of $C_\beta 2$. The 6.5-kb band contains 3' $C_\beta 2$ sequences. Rearrangements involving $C_\beta 2$ are detected as a loss of the 8.0-kb band and the appearance of a new band. Rearrangements of $C_\beta 1$ are inferred by the loss of the 3.5-kb band.

Expression of TcR β results in a 1.3- or 1.0-kb mRNA (Yanagi *et al.*, 1984). The 1.3-kb message is the product of a complete VDJ rearrangement, while the 1.0-kb message may be either the result of transcription of DJ rearrangements or from transcription beginning 5' of a J. Whether the transcript contains $C_\beta 1$ or $C_\beta 2$ may be determined by using a probe to the 3' untranslated region of the gene (Yoshikai *et al.*, 1984).

3.3. Analysis of γ-Chain Structure and Expression

A genomic probe to the J region of the γ gene detects two bands of 15 and 12.5 kb in DNA digested with Bam H1. The 15-kb band contains the $_\gamma 1$ sequences while the 12.5-kb band contains the $_\gamma 2$ sequences. DNA cleaved with Eco R1 and probed with a J_γ fragment gives 2 bands of 3.3 kb and 1.8 kb. The 3.3 kb region contains $J_\gamma 2$ sequences, while the 1.8 kb region contains $J_\gamma 1$ sequences. The J_γ probe will also detect the same 15-kb and 12.5-kb Bam H1 fragments as are detected by the constant region probe.

Unlike the α and β genes, only a single distinct γ transcript of 1.6 kb is made (LeFranc *et al.*, 1986a,b).

4. Rearrangement and Expression of the TcR Genes in Human Neoplasms and Related Disorders

In cells of the B lymphoid lineage, the rearrangement of the Ig genes has proven to be a useful marker of clonality (Korsmeyer *et al.*, 1983). Prior to the advent of probes to the genes of the T-cell receptor there were no good markers of clonality in the T-cell lineage. Since the cloning of the TcR genes numerous malignant and nonmalignant samples from patients with hematologic and other diseases have been examined. This has made it possible to determine whether a population of T cells is clonal or polyclonal in origin. The results of such studies are summarized below.

5. T-Cell Malignancies

5.1. Acute T-Lymphoid Neoplasms

5.1.1. T-Cell Acute Lymphoblastic Leukemia (T-ALL)

T-ALL cells and T-cell leukemic cell lines have been assessed for the structure and expression of the α-, β-, and γ-chain genes. Clonal TcR$_\beta$ rearrangement of one or both β-chain alleles has been identified in almost all of the reported cases of T-ALL (Minden *et al.*, 1985; Rabbitts *et al.*, 1985a; Tawa *et al.*, 1985; O'Connor *et al.*, 1985; Bertness *et al.*, 1985; Waldmann *et al.*, 1985; Pelicci *et al.*, 1985). There is no predominant rearrangement seen among these tumors indicating that rearrangement is a random event in the cells of this disease. Also there is no correlation between the structure of TcR$_\beta$ and T-cell surface antigens such as T1, T3, T4, T6, T8, and T10. However expression of the cell surface antigen Leu 9 appears to correlate with rearrangement of TcR$_\beta$ (Prittaluga *et al.*, 1986).

In all T-ALLs in which the β-chain gene was rearranged the γ-chain gene was also rearranged (Greenberg *et al.*, 1986; Lefranc *et al.*, 1986a; Lefranc and Rabbitts, 1985; Quertermous *et al.*, 1986).

A number of cases have been studied at the time of presentation, while in remission, and at the time of relapse. In the few cases that have been reported the pattern of rearrangement at presentation and in relapse was identical, indicating that relapse was due to reemergence of the original malignant clone. This would indicate that the accepted definition of remission, that is the presence of normal peripheral blood counts and a normal marrow with less than 5% blast cells is inadequate for detecting residual leukemic cells. Using the presence of a clonal rearrangement as a marker of the malignant clone may improve the detection of leukemic cells. The sensitivity of this technique for detecting residual disease is dependent upon a number of factors including the specific activity of the probe, the efficiency of the transfer and the retention of DNA on the supporting membrane. By optimizing these conditions it may be possible to detect a clone of malignant cells at the 0.2% level (Zehnbauer *et al.*, 1986). However if the sensitivity of the technique becomes too great one may lose the ability to distinguish rearrangements that have occurred in normal T cells from those that mark the malignant population of T cells. Regardless of these constraints, this adjunct to morphologic assessment of the bone marrow during remission may prove useful in determining the type and timing of therapy that a patient should receive. However, before this may be used clinically studies indicating the consequence of detecting residual disease are needed.

5.1.2. Acute T-Cell Leukemia/Lymphoma

Acute T-cell leukemia/lymphoma is a T-cell malignancy that is found predominantly in Southern Japan and the Caribbean (Jarrett *et al.*, 1986). The disease is characterized by lymphadenopathy, skin infiltrates, bony lesions, and hypercalcemia. The disease tends to take an aggressive course. The leukemic cells have a mature phenotype expressing T3, T4, IL-2 receptor, and Ia on the cell surface. The retrovirus, HTLV-1, has been implicated in the etiology of the disease since it is found in the leukemic cells of a large number of these patients.

Studies of TcR$_\beta$ and γ in patients with this disease have revealed clonal rearrangements of these genes (Jarrett *et al.*, 1986; Davey *et al.*, 1986). As in T-ALL no predominant pattern of rearrangement was seen when comparing the different cases. In addition to the existence of a clonal pattern of TcR rearrangement, the HTLV-1 virus is integrated in a clonal fashion.

5.1.3. T-Cell Prolymphocytic Leukemia

This disease is characterized by a high white cell count, splenomegaly, and an aggressive clinical course with poor response to therapy. In all cases that have been analyzed, TcR$_\beta$ has been rearranged (Minden *et al.*, 1985; O'Connor *et al.*, 1985; Rabbitts *et al.*, 1985a; Pelicci *et al.*, 1985; Baer *et al.*, 1985).

5.2. Chronic T-Cell Malignancies

5.2.1. Sezary Syndrome/Mycoses Fungoides

These diseases likely represent a spectrum of the same disease with mycoses fungoides being the tissue phase of the disease and Sezary syndrome representing the leukemic phase of the disease. Mycoses fungoides (MF) is characterized by erythroderma, cutaneous infiltrates, and lymphadenopathy. The nuclei of the cells have a typical cerebriform shape. The cell surface phenotype is T3+, T4+, T8−. In all cases of this disease that have been studied, TcR$_\beta$ has been rearranged (Rabbitts *et al.*, 1985a; Flug *et al.*, 1985; O'Connor *et al.*, 1985; Aisenberg *et al.*, 1985; Bertness *et al.*, 1985; Waldmann *et al.*, 1985; Pelicci *et al.*, 1985; Weiss *et al.*, 1985a). In only a few cases has concomitant rearrangement of Ig HC been detected.

In approximately half of the cases of MF there is significant lymphadenopathy. In affected lymph nodes malignant cells are not usually observed histopathologically and so it was concluded that this dermatopathic lymphadenopathy was a reaction to the malignant process that occurred in the skin or the result of chronic infection of the skin. However, molecular analysis of these nodes has revealed that the same rearrangement pattern that is detected in the skin can be detected in cells derived from the lymph node. Thus the development of lymphadenopathy with its associated poorer prognosis correlates with an increased tumor burden (Weiss *et al.*, 1985a).

5.2.2. T-Cell Chronic Lymphocytic Leukemia

T-CLL tends to follow a benign clinical course, with prolonged survival and a relatively low white blood cell count. Over time, in some cases, the disease becomes more aggressive. The prolonged course has led to the suggestion that this may not be a neoplastic disease. Regardless of this, the disease has been shown to be clonal in nature by the discovery of TcR β-chain rearrangements in all of the cases that have been evaluated to date (Rabbitts *et al.*, 1985a; Flug *et al.*, 1985; Pelicci *et al.*, 1985; Aisenberg *et al.*, 1985; O'Connor *et al.*, 1985; Foa *et al.*, 1986). In a few cases concomitant rearrangement of Ig heavy chain (HC) was noted.

5.3. T8 Lymphocytosis

This disorder is felt by some to be a variant of T-CLL. Patients present with splenomegaly, peripheral blood and bone lymphocytosis, granulocytopenia, and recurrent infections. A few cases with red-cell aplasia have been reported. Approximately one third of patients with T-γ lymphocytosis have a prior history of rheumatoid arthritis suggesting an association between the two diseases.

The lymphocytes which are in excess in this disease have the characteristics of natural killer cells (NK). In a number of cases the cells have the morphological appearance of large granular lymphocytes, express cell surface markers characteristic of NK cells, have high antibody-dependent complement lysis activity, and occasionally express spontaneous NK activity.

Natural killer cells are believed to makeup approximately 10% of the circulating lymphocytes in normal individuals. These cells are T11+, T3−, T8+. Cell lines with NK-like activity and cell surface markers have been established. Results of molecular studies of such cell lines has been controversial. In some cases $TcR_β$ rearrangement and a $TcR_β$ mRNA has been found while other studies have been negative (Lanier et al., 1986). It remains to be determined whether the cell types expressing NK activity but differing in the structure/expression of their TcR genes, represent different stages of the same lineage or whether they are distinctly different cell types that have in common NK-like activity.

Studies of patients with T8 lymphocytosis has given equally confusing results. In some cases, polyclonal rearrangements of $TcR_β$ have been detected, while in others, a clonal pattern was seen indicating that in some patients this is a clonal disease (Aisenberg et al., 1985; Bertness et al., 1985; Greenberg et al., 1986; Berliner et al., 1986; Rambaldi et al., 1985); but this also raises a number of questions. Does the disease begin as a clonal disease or does it begin as a polyclonal disease that over time progresses to a clonal state? Are the cells involved in the clonal disease the same as in the polyclonal disease? Is this a malignant state? Answers to these questions will require long-term longitudinal studies and more detailed analysis of the involved cells.

5.4. T-Cell Lymphomas

The morphologic discrimination between lymphomas of the T- and B-cell lineage is difficult. This has been improved by the application of immunohistochemistry. In the case of the B-cell neoplasms the presence of a monoclonal population of cells expressing either κ or λ immunoglobulin light chains (LC) has proven useful both in establishing the presence of a clonal population of cells and in defining the malignant cells as being of the B-cell lineage. Occasionally, though, the lymph node may contain a large percentage of T cells and a clonal population of B cells is not apparent by immunophenotyping. The clonal nature and the cellular lineage of these cases are ambiguous.

The application of molecular probes derived from the Ig and TcR genes has provided some clarification in such cases. DNA from lymph node specimens containing a large number of T lymphocytes have been examined for rearrangement of their immunoglobulin and TcR genes. In the majority of cases, rearrangement of the TcR genes was found (Rabbitts *et al.*, 1985a; Flug *et al.*, 1985; O'Connor *et al.*, 1985; Isaacson *et al.*, 1985; Bertness *et al.*, 1985; Waldmann *et al.*, 1985; Pelicci *et al.*, 1985; Griesser *et al.*, 1986a,b). In other cases, rearrangement of Ig HC was found without rearrangement of TcR. However, in a few cases, rearrangement of both genes was found (Pelicci *et al.*, 1985; Griesser *et al.*, 1986a,b). These results are interpreted as follows: (1) The cases in which there is only TcR rearrangement without Ig rearrangement are T-cell lymphomas. (2) The cases in which only Ig rearrangement was found despite a predominance of T cells represent B-cell lymphomas with infiltration of normal T cells. (3) Cases in which both Ig and TcR rearrangements occur require further evaluation, as to the structure of the immunoglobulin light-chain in genes and cell surface phenotype in order to assign cell lineage.

5.5. Non-B Non-T ALL

The majority of childhood ALLs are of the so called non-B non-T phenotype. This name was given to this group of diseases because the cells do not express immunoglobulin proteins either in their cytoplasm or their cell membranes (a defining characteristic of B cells) and do not form rosettes with sheep red blood cells (a defining characteristic of mature T cells). More recent characterization of these cells has revealed that they often express the B-cell surface antigens B-1 and B-4 and the lymphoid antigens cALLa and TdT. Almost all of the cases that have been studied have Ig HC rearrangement (Linch *et al.*, 1985). Based on these results it was felt that non-B non-T ALLs in fact represent a cell at an early stage of B-cell development. Characterization of these cells with TcR probes however revealed up to 30% of cases of non-B non-T ALL have rearrangement of their TcR β-chain genes (Pelicci *et al.*, 1985; Minden *et al.*, 1985; Tawa *et al.*, 1985) and up to 50% have rearrangement of the γ-chain genes (Greenberg *et al.*, 1986). Studies correlating clinical outcome with the different genetic subgroups have not yet been presented. These findings again have led to confusion as to the lineage of the leukemic cells.

5.6. B-CLL

B-CLL is characterized by the overproduction of small lymphocytes. In the majority of cases, the cells express immunoglobulin as well as other B-cell antigens on their cell surfaces. In such cases rearrangement of the immunoglobulin light- and heavy-chain genes is found; up to 10% of cases also may have rearranged their TcR β genes (O'Connor *et al.*, 1985; Waldmann *et al.*, 1985; Pelicci *et al.*, 1985).

5.7. B-Cell Lymphoma

Five of 42 cases in which both Ig HC and TcRβ gene rearrangement was assessed had a rearrangement of both loci, while the remainder had rearrangement of the Ig HC genes only.

5.8. Lymphoproliferative Disorders of Uncertain Origin

There are a number of diseases of the lymphoreticular system that are characterized by lymphadenopathy with or without splenomegaly such as Lennert's lymphoma, angioimmunoblastic lymphadenopathy (AIL), Ki 1+ lymphoma, true histiocytic lymphoma, and Hodgkin's lymphoma. Studies of the structure of the immunoglobulin and TcR genes in such cases have identified clonal populations in most of these malignant diseases and, in some cases, have indicated the lineage from which the malignant clone originated.

5.8.1. Lennert's Lymphoma

The pathology of the lymph node reveals large epithelioid cells. The cell surface markers are generally those of T cells. In 6 of 6 cases the TcR β chain locus was rearranged while the immunoglobulin genes were in the germline configuration. This indicates that Lennert's lymphoma is a clonal disease that is of likely T-cell origin (O'Connor et al., 1985; Griesser et al., 1986a,b; Feller et al., 1986a,b).

5.8.2. Angioimmunoblastic Lymphadenopathy (AIL)

This is a disease characterized by an indolent course punctuated by generalized arthralgia, lymphadenopathy, frequent infections, and evolving over time to an aggressive disease with cutaneous lymphoid infiltrates. The lesions contain T cells of varied morphology with admixed polyclonal plasmacytes and eosinophils, loss of germinal centers, and infiltration by arborizing vessels. A variant of this exists in which the pathology is similar except that the germinal centers persist; this has been referred to as hyperimmune response (HR). HR is a generally benign disease but may progress to AIL. Nine cases of AIL and two cases of HR were studied for gene rearrangement. All nine cases of AIL revealed TcR rearrangement while rearrangement was not found in the two cases of HR (Griesser et al., 1986a,b).

This indicates that AIL is a clonal disease likely of T-cell origin. The lack of rearrangement in the two cases of HR may be due to one of the following three models: (1) the contamination of the sample by the residual normal cells obscures the presence of the malignant clone; (2) HR and AIL are two different diseases; and (3) HR is a polyclonal disease that over time evolves into the monoclonal disease AIL. In support of models one and three is the case report of a patient who was studied at two different times (Bertness et al., 1985). The initial biopsy was negative for rearrangement while the later biopsy showed new bands. In order to resolve these questions, cell-sorting experiments and longitudinal studies will be necessary.

5.8.3. Non-Hodgkin's Lymphoma Containing Large Anaplastic Ki 1+ Cells

A population of lymphomas has been defined by the use of a monoclonal antibody, Ki 1, which detects a cell surface antigen found on Reed-Sternberg cells, activated normal T lymphocytes, and HLA DR+ lymphoma cells in some large anaplastic lymphomas. Immunophenotypic studies of these tumors suggests that they are of the T lineage. In a study of ten Ki 1+ lymphomas, nine of which had a T-cell marker phenotype, clonal rearrangement of TcR_β was found in seven cases. Three of the seven cases also had Ig HC gene rearrangement. One case had cell surface markers compatible with the B-cell lineage; rearrangements of both Ig HC and TcR_β were found in this case (Griesser *et al.*, 1986a). Two of the three cases that had germline TcR_β and Ig HC genes had rearranged γ-chain genes (Griesser *et al.*, 1986a).

5.8.4. Histiocytic Lymphoma

The histiocytic malignancies are a subset of lymphomas in which the origin of the malignant cell is unclear. Morphologically the cells resemble histiocytes. Immunophenotyping of these cells has not been helpful in that they often lack the markers of B or T cells. In a study of six such cases it was found that TcR_β was rearranged in four cases, Ig HC and TcR_β was rearranged in one case, and in another case both genes were in the germline configuration (Weiss *et al.*, 1985b). The structure of the γ-chain locus was not assessed. These findings suggest that most cases of histiocytic lymphoma are of T-cell origin. The one case that did not show rearrangement of either Ig or TcR genes may represent a true histiocytic lymphoma, however in the absence of the study of the γ chain or the use of a more extensive panel of myeloid markers it is difficult to assign the tumor to a given lineage.

5.8.5. Lymphomatoid Papulosis

This is a chronic disease characterized by the episodic development of papular skin lesions that ulcerate and then spontaneously heal by scarring. Histologically, the lesions appear to be either malignant lymphomas containing large atypical cells with cerebriform nuclei or Reed-Sternberg-type cells. Lesions from five of six patients showed clonal rearrangements of both TcR β and γ. In the one case that did not show rearrangement, the infiltrate was small and likely did not contribute significantly to the sample. Two cases require special mention. In one case, three rearranged bands were seen in a single lesion and, in another case, lesions from different sites contained different patterns of rearrangement. These observations suggest that more than one clone may be present in one individual (Weiss *et al.*, 1986b).

5.8.6. Hodgkin's Disease

The clinical course of Hodgkin's disease would indicate that it is a malignant state. However, attempts at identifying a clonal population in affected lymph nodes has been generally unsuccessful. The inability to detect a malignant clone of cells may be due to the fact that, in addition to the malignant cells, the lymph node is infiltrated by very large numbers of normal reactive cells. Some evidence that Hodgkin's disease is a clonal disease comes from studies of cell lines derived from patients with late-stage disease. The structure of the immunoglobulin and TcR$_\beta$ genes were studied in a number of cell lines. In some lines, rearrangement of the immunoglobulin genes was found, while in one line only rearrangement of TcR$_\beta$ was found (Rabbitts et al., 1985a; Linch et al., 1985).

In an initial study of eight patients clonal rearrangement of TcR$_\beta$ and TcR$_\gamma$ was found in four cases. The intensity of the rearranged bands was of a degree that suggested only a small fraction of the cells were part of the clone (Griesser et al., 1986b). A similar observation was made by Knowles et al. (1986). These findings suggest that Hodgkin's disease is in fact a clonal disorder involving T cells. The clonal T cells may be malignant, however another explanation must be considered. The lymph nodes in Hodgkin's disease are filled with reactive cells. It is possible that the clonal rearrangement that is detected is due to the expansion of a single clone of cells not as a result of their malignant transformation, but as a response to a specific antigen.

A different situation is found in some cases of Hodgkin's disease in which the proliferation of Reed-Sternberg cells is prominent. In 5 of 7 of these cases clonal rearrangement of the immunoglobulin genes was found (Weiss et al., 1986a). In six similar cases, γ-chain rearrangement was found in two cases and Ig HC rearrangement was found in two cases (H. Griesser, A. Feller, T.W. Mak, and K. Lennert, submitted). Assuming that the clonal Ig rearrangements derive from the Reed-Sternberg cells, the evidence favors the view that some Reed-Sternberg cells have B-cell character while others resemble pre-T cells.

Regardless of these questions, the finding of a clonal population of cells in Hodgkin's disease has opened the way for investigations that should shed light on the etiology of this disease (Griesser et al., 1986b; Knowles et al., 1986).

5.9. Another Hematopoietic Malignancy

Acute myeloblastic leukemia. Acute myeloblastic leukemia (AML) is a malignancy involving a cell that is the progenitor of granulocytes, erythrocytes, and platelets. In a survey of 24 cases of AML (all of which expressed myeloid cell surface markers), 3 cases were found with rearrangement of TcR$_\beta$; one case also had rearrangement of the Ig HC gene (Cheng et al., 1985).

6. Observations

6.1. Rearrangement as a Means of Defining Lineage

Based on the data presented in this review and elsewhere, a number of conclusions may be made concerning the use of information derived from the investigation of the structure of the TcR genes or the immunoglobulin genes for the definition of cell lineage. By and large, rearrangement of the Ig genes or the TcR$_\beta$ chain locus indicates that the expanded population of cells is of the B-cell or T-cell lineage respectively. However, there are a couple of notable exceptions: (1) Those cases in which both types of genes are rearranged, and (2) cases of AML that have rearranged either their Ig or TcR$_\beta$ genes. Because of these exceptions, the assignment of lineage must take into account all available information, including the structure of the Ig LC genes, and the immunologic phenotype of the cell.

Rearrangements of the γ-chain genes are useful as clonal markers in AML, B-cell, and T-cell malignancies, however, since they are found not infrequently in each of these diseases they are not useful for assigning lineage. In general all cases that have rearranged TcR β have rearranged γ [a few exceptions to this have recently been observed (T.W. Mak, unpublished observation)] (Griesser *et al.*, 1986a; Quertermous *et al.*, 1986; Greenberg *et al.*, 1986; Ohashi *et al.*, 1985; Davey *et al.*, 1986). There are a few cases reported in which only the γ chain is rearranged (Griesser *et al.*, 1986b). Such cells may represent a very early stage of T-lymphocyte development, alternatively they may define a previously unrecognized type of immune response cell. Studies of such tumors and normal cells that are T3 positive but lack TcR$_\beta$ rearrangements will clarify this issue (Brenner *et al.*, 1986).

As mentioned previously, the evaluation of either rearrangements or expression of the α chain is not at a point of refinement that it can be used with confidence in assigning cell lineage.

7. T-Cell Ontogeny

In Chapter 5 the ontogeny of T cells was discussed. Studies of cells within the thymus indicate that rearrangement and expression occurs in a hierarchical manner with TcR$_\gamma$ occurring first followed by TcR$_\beta$ and finally TcR$_\alpha$ (Raulet *et al.*, 1985; Snodgrass *et al.*, 1985). Since leukemic cells are likely arrested at a stage along their developmental pathway, their study has been useful in identifying and confirming developmental programs. The earliest T cell is recognized by the expression of the cell surface antigen 3A1 (Leu 9) and the absence of other markers such as T 11. Leukemic cells of this phenotype have been found to have TcR$_\gamma$ and TcR$_\beta$ rearrangements (Davey *et al.*, 1986).

The earliest T-cell precursors express TcR_β and TcR_γ mRNA while more mature cells express TcR_β and TcR_α and very little to no TcR_γ mRNA (Davey *et al.*, 1986). The expression of TcR_α is found in cells that express T3. The expression of the T3 antigen on the cell surface requires that the cell produces either functional TcR_β and TcR_α proteins (Acuto and Reinherz, 1985; Borst *et al.*, 1983; Ohashi *et al.*, 1985) or functional TcR_γ and a yet unidentified other protein (Brenner *et al.*, 1986).

7.1. Lymphoid Malignancies with Rearrangement of Both TcR and Ig Genes

Rearrangements of TcR and Ig genes have been found in approximately 10% of T-cell neoplasms and up to 30% of B and pre-B cell neoplasms. Rearrangements of both loci have been found in cloned cell lines indicating that in some cases this finding is due to both genes being arranged in the same cell and not due to two coexisting cell populations. The machinery required to rearrange genes and the recognition sequences bounding the J genes are very similar in both Ig and TcR families (Yancopoulos *et al.*, 1986). It is possible that, during the early stages of B-cell and T-cell development that the process of rearrangement lacks fidelity so that a cell that is destined to be a B cell may undergo rearrangement of the TcR_β chain while a cell that is destined to be a T cell may undergo a rearrangement of Ig HC genes and that rearrangement of either TcR_β or Ig HC loci does not commit a cell to one lineage or the other (Greaves *et al.*, 1986). Of note, however, is that the majority of mature B cells have not rearranged their TcR_β genes. An alternative explanation is that rather than reflecting a random event in early lymphoid development concomitant rearrangement of TcR_β and Ig HC genes results from abnormal gene expression in the malignant cell. As such, the finding of both rearrangements in the same cell is an example of lineage infidelity (Smith *et al.*, 1983).

7.2. T-Cell Malignancies with More than Two Rearranged Bands

Another interesting observation is that, in some cases, more than two bands are rearranged in the same sample. This may either represent multiple rearrangements within a single cell or may be due to the presence of more than one population of malignant cells. In support of the former hypothesis is the finding that in cloned T-cell lines more than two rearranged bands may be seen (Sangster *et al.*, 1986). The multiple bands may come about as a result of sister chromatid exchange, chromosomal duplication, or reintegration of a piece of spliced out DNA (Kronenberg *et al.*, 1986). In support of the possibility of there being more than one clone is the finding in lymphomatoid papulosis of different patterns of rearrangement in lesions from different sites (Weiss *et al.*, 1986b). This is analogous to the finding of a number of clones, in the same individual, with B-cell malignancies (Sklar *et al.*, 1984). To resolve this question it will be necessary to isolate the genetic information of an individual cell by either cellular cloning of single cells or by immortalizing an individual cell by hybridizing it to a non-T cell and then studying the structure of the TcR genes in the hybrid cell lines. If multiple bands are the result of the presence of more than one clonal populations of cells, one will find a reduced number of bands in the hybrid, while, if the multiple bands all exist in a single cell, the hybrid cell will have the same pattern of rearrangement.

7.3. Involvement of TcR Genes in Chromosome Translocations

Consistent chromosomal aberrations have been recognized in a variety of leukemias and are felt to play a role in the development of the malignant phenotype (Klein, 1983; Rowley, 1978; Croce *et al.*, 1984). In another chapter of this book the involvement of the TcR genes in chromosome translocations is reviewed in detail. Briefly, a number of consistent chromosome translocations have been found in T-cell malignancies that directly involve the α-chain locus or are close to the sites of the β- or γ-chain loci (Baer *et al.*, 1985; Bakhshi *et al.*, 1985; Hecht *et al.*, 1985; Hecht *et al.*, 1984; Lewis *et al.*, 1985; Rabbitts *et al.*, 1985b; Sadamori *et al.*, 1985; Shima *et al.*, 1986; Smith and Kirsch, n.d.; Smith *et al.*, 1986; Williams *et al.*, 1984; Zech *et al.*, 1984). In two cases the breakpoints have been isolated. In one instance the α-chain genes had rearranged with the immunoglobulin heavy chain genes (Denny *et al.*, 1986a,b), while in the other instance, the α-chain locus had rearranged with the *c-myc* oncogene on chromosome 8 (Shima *et al.*, 1986); this provides evidence that in certain cases the translocation maybe involved in the development of the malignant state. Whether the other translocations also involve oncogenes is being pursued in many laboratories.

8. Conclusion

Rearrangements of the TcR loci have proven useful in determining clonality, in aiding in the assignment of cell lineage, and in following the progression of various lymphoproliferative diseases. Such rearrangements, although necessary for T lineage commitment, are not entirely specific.

The detection of rearrangements may be useful in defining prognosis. For example, do patients whose cells have only TcR α-, β- and γ-chain rearrangements, but not Ig HC rearrangements, have a different prognosis than patients whose cells have rearrangements of TcR_β and Ig HC? Do patients whose cells have rearranged TcR_β but not TcR_α have a different prognosis from patients whose cells have also rearranged TcR_α?

With Southern blot analysis it is possible to detect very small numbers of malignant cells. Will the detection of persistent minimal residual disease in what appears to be morphological complete remission improve our understanding of the relationship that exists between the normal and leukemic cells? Will the detection of a residual malignant clone predict leukemic relapse? Will early treatment or more aggressive induction therapy of patients showing residual leukemia while in morphologic complete remission improve the prognosis of this group of patients? The probes that are now available may be used to address such questions and in the future such analyses are likely to affect the clinical practice of hematology and oncology.

References

Acuto, O., and Reinherz, E.L., 1985, The human T cell receptor: Structure and function, *N. Eng. J. Med.* **312:**1100–1111.

Aisenberg, A.C., Krontiris, T.G., Mak, T.W., and Wilkes, B.M., 1985, Rearrangement of the gene for the beta chain of the T cell receptor in T cell chronic lymphocytic leukemias and related disorders. *N. Eng. J. Med.* **313:**529–533.

Arnold, A., Cosman, J., Bakhshi, A., Jaffe, E.S., Waldmann, T.A., and Korsmeyer, S.J., 1983, Immunoglobulin gene rearrangements as unique clonal markers in human lymphoid neoplasms, *N. Eng. J. Med.* **309:**1593–1599.

Baer, R., Chen, K.-C., Smith, S.D., and Rabbitts, T.H., 1985, Fusion of an immunoglobulin variable gene and a T cell receptor constant gene in the chromosome 14 inversion associated with T cell tumors, *Cell* **43:**705–713.

Baer, R., Lefranc, M.-P., Minowada, J., Forster, A., Stinson, M.A., and Rabbitts, T.H., 1986, Organization of the T cell receptor alpha chain gene and rearrangements in human T cell leukemias, *Mol. Biol. Med.* **3:**265–277.

Bakhshi, A., Jensen, J.P., Boldman, P., Wright, J.J., McBride, O.W., Epstein, A.L., and Korsmeyer, S.J., 1985, Cloning the chromosomal breakpoint t(14;18) human lymphomas: Clustering around JH on chromosome 14 and near a transcriptional unit on 18, *Cell* **41:**899–906.

Bank I., DePinho, R.A., Brenner, M.B., Cassimeris, J., Alt, F.W., and Chess, L., 1986, A functional T3 molecule associated with a novel heterodimer on the surface of immature thymocytes, *Nature* **322:**179–181.

Berliner, N., Duby, A.D., Linch, D.C., Murre, C., Quertermous, T., Knott, L.J., Azin, T., Newland, A.C., Lewis, D.L., Galvin, M.C., and Seidman, J.G., 1986, T cell receptor gene rearrangements define a monoclonal T cell proliferation in patients with T cell lymphocytosis and cytopenia, *Blood* **67:**914–918.

Bertness, V., Kirsch, I., Hollis, G., Johnson B., and Bunn, P.A., Jr., 1985, T cell receptor gene rearrangements as clinical markers of human T cell lymphomas, *N. Eng. J. Med.* **313:**534–538.

Borst, J., Alexander, S., Elder, J., and Terhorst, C., 1983, The T3 complex on human T lymphocytes involves four structurally distinct glycoproteins, *J. Biol. Chem.* **258:**5135–5143.

Brenner, M.B., McLean, J., Dialynas, D.P., Strominger, J.L., Smith, J.A., Owen, F.L., Seidman, J.G., Ip, S., Rosen, R., and Krangel, M.S., 1986, Identification of a putative section T cell receptor, *Nature* **322:**145–149.

Caccia, N., Bruns, G., Kirsch, I., Hollis, G.F., Bertness, A.D., and Mak, T.W., 1985, T cell receptor alpha chain genes are located on chromosome 14 at 14q11–14q12 in humans, *J. Exp. Med.* **161:**1255–1260.

Caccia, N., Kronenberg, M., Saxe, D., Haars, R., Bruns, G., Goverman, J., Malissen, M., Willard, H., Yoshikai, Y., Simon, M., Hood, L., and Mak, T.W., 1984, The T cell receptor beta chain genes are located on chromosome 6 in mice and chromosome 7 in humans, *Cell* **37**:1091–1099.

Cheng, G., Minden, M.D., Mak, T.W., and McCulloch, E.A., 1985, T cell receptor and immunoglobulin gene rearrangements in acute myeloblastic luekemia: Evidence for lineage infidelity, *J. Exp. Med.* **163**:414–424.

Clark, S.P., Yoshikai, Y., Siu, G., Taylor, S., Hood, L., and Mak, T.W., 1984, Identification of a diversity segment of the human T cell receptor beta chain, and comparison to the analogous murine element, *Nature* **311**:387–389.

Collins, M.K.L., Goodfellow, P.N., Spurr, N.K., Solomon, E., Tanigawa, G., Tonegawa, S., and Owen, M.J., 1985, The human T cell receptor alpha chain gene maps to chromosome 14, *Nature* **314**:273–274.

Croce, C.M., Isobe, M., Palumbo, A., Puck, J., Ming., J., Tweardy, D., Erikson, J., Davis, M., and Rovera, G., 1985, Gene for alpha chain of human T cell receptor: Location on chromosome 14 region involved in T cell neoplasms, *Science* **227**:1044–1047.

Croce, D.M., Tusjimoto, Y., Erikson, J., and Nowell, P., 1984, Biology of disease: Chromosome translocations and B cell neoplasia, *Lab. Invest.* **51**:258–267.

Davey, M.P., Bongiovanni, K.F., Kaulfersch, W., Quertermous, T., Seidman, J.G., Hershfield, M.S., Kurtzberg, J., Haynes, B.F., Davis, M.M., and Waldmann, T., 1986, Immunoglobulin and T cell receptor gene rearrangement and expression in human lymphoid leukemia cells at different stages of maturation, *Proc. Natl. Acad. Sci. USA* **83**:8759–8763.

Denny, C.T., Hollis, G.F., Hecht, F., Morgan, R., Link, M.P., Smith, S.D., and Kirsch, I.R., 1986, Common mechanism of chromosome inversion in B- and T-cell tumors: relevance to lymphoid development, *Science* **234**:197–200.

Denny, C.T., Yoshikai, Y., Mak, T.W., Smithe, S.D., Hollis, G.F., and Kirsch, I.R., 1986, A chromosome 14 inversion in a T cell lymphoma is caused by the site specific recombination between immunoglobulin and T cell receptor loci, *Nature* **320**:549–551.

Feller, A.C., Grieser, G.H., Mak, T.W., and Lennert, K., 1986, Lymphoepithelioid lymphoma, (Lennert's lymphoma) is a monoclonal proliferation of helper/inducer T cells, *Blood* **68**:663–667.

Fialkow, P.J., 1976, Clonal origin of human tumors, *Biochem. Biophys. Acta* **458**:283–321.

Flug, F., Pelicci, P.-G., Bonetti, F., Knowles, D.M., and Dalla-Favera, R., 1985, T cell receptor gene rearrangements as markers of lineage and clonality in T cell neoplasms, *Proc. Natl. Acad. Sci. USA* **82**:3460–3464.

Foa, R., Pellici, P.G., Migone, N., Lauria, F., Pizzolo, G., Flug, F., Knowles, D.M., and Dalla-Favera, R., 1986, Analysis of T cell receptor beta chain gene rearrangements demonstrates the monoclonal nature of T cell chronic lymphoproliferative disorders, *Blood* **67**:247–250.

Furley, A.J., Mixutani, S., Weilbaecher, K., Dhaliwal, H.S., Bord, A.M., Chan, L.C., Molgaard, H.V., Toyonaga, B., Mak, T., van den Elsen, P., Gold, D., Terhorst, C., and Greaves, M.F., 1986, Developmentally regulated rearrangement and expression of genes encoding the T cell receptor T3 complex, *Cell* **46**:75–87.

Greaves, M.F., Chan, L.C., Furley, A.J.W., Watt, S.M., Molgaard, H.V., 1986, Lineage promiscuity in hematopoietic differentiation and leukemia, *Blood* **67**:1–11.

Greenberg, J.M., Quertermous, T., Seidman, J.G., and Kersey, J.H., 1986, Human T cell gamma chain gene rearrangements in acute lymphoid and nonlymphoid leukemia: comparison with T cell receptor beta chain gene, *J. Immunol.* **137**(6):2043–2049.

Griesser, H., Feller, A., Lennert, K., Tweedale, M., Messner, H.A., Zalcberg, J., Minden, M.D., and Mak, T.W., 1986a, The structure of the T cell gamma chain gene in lymphoproliferative disorders and lymphoma cell lines, *Blood* **68**:592–594.

Griesser, H., Feller, A., Lennert, K., Minden, M.D., and Mak, T.W., 1986b, Rearrangement of the beta chain of the T cell antigen receptor and immunoglobulin gene in lymphoproliferative disorders, *J. Clin. Invest.* **78:**1179–1184.

Hecht, F., Morgan, R., Gemill, R.M., Hecht, B., and Smith, S.D., 1985, Translocations in T cell leukemia and lymphoma, *N. Eng. J. Med.* **313:**758–759.

Hecht, F., Morgan, R., Hecht, R., and Smith, S.D., 1984, Common region on chromosome 14 in T cell leukemia and lymphoma, *Science* **226:**1445–.

Hedrick, S.M., Cohen, D.I., Nielsen, E.A., and Davis, M.M., 1984, Isolation of cDNA clones encoding T cell specific membrane associated proteins, *Nature* **309:**149–153.

Hedrick, S.M., Nielsen, E.A., Kavaler, J., Cohen, D.I., and Davis, M.M., 1984b, Sequence relationships between putative T cell receptors polypeptides and immunoglobulins, *Nature* **309:**153–158.

Isaacson, P.G., Spencer, J., Connolly, C.E., Pollock, D.J., Stein, H., O'Connor, N.T.J., Bevan, D.H., Kirkham, N., Wainsoat, J.S., and Mason, D.Y., 1985, Malignant histiocytosis of the intestine: A T cell lymphoma, *Lancet* **ii:**688–691.

Isobe, M., Erikson, J., Emanuel, B.S., Nowell, P.C., and Croce, C.M., 1985, Location of gene for beta subunit of human T cell receptor at band 7q35, a region of rearrangements in T cells, *Science* **228:**580–582.

Jarrett, R.F., Mitsuya, H., Mann, D.L., Cossman, J., Broder, S., and Reitz, M.S., 1986, Configuration and expression of the T cell receptor beta chain gene in human T lymphotropic virus I infected cells, *J. Exp. Med.* **163:**383–399.

Kappler, J., Kubo, R., Haskins, K., Hannum, C., Marrack, P., Pigeon, M., McIntyre, B., Allison, J., and Trowbridge, I., 1983, The major histocompatibility complex restricted antigen receptor on T cells in mouse and man: Identification of constant and variable peptides, *Cell* **35:**295–302.

Kimura, H., Toyanaga, B., Yoshikai, Y., Triebel, F., Debre, P., Minden, M., Mak, T.W., 1986, Sequence and diversity of the human T cell beta chain V region gene, *J. Exp. Med.* **164:**739–750.

Klein, G., 1983, Specific chromosomal translocations and the genesis of B cell derived tumors in mice and men, *Cell* **32:**311–315.

Knowles, D.M., Neri, A., Pellici, A.N., Burke, J.S., Wu, A., Winberg, C.D., Sheibani, K., and Dalla-Favera, R., 1986, Immunoglobulin and T cell receptor beta chain gene rearrangement analysis of Hodgkin's disease: implications for lineage determination and differential diagnosis, *Proc. Natl. Acad. Sci. USA* **83:**7942–7946.

Korsmeyer, S.J., Arnold, A., Bakshi, A., Ravetch, J.V., Siebenlist, U., Hieter, P.A., Sharrow, S.O., LeBien, T.W., Kersey, J.H., Poplack, D.G., Leder, P., and Waldmann, T.A.,, 1983, Immunoglobulin gene rearrangement and cell surface antigen expression in acute lymphocytic leukemia of T cell and B cell precursor origins, *J. Clin. Invest.* **71:**301–313.

Kronenberg, M., Siu, G., Hood, L.E., and Shastri, N., 1986, The molecular genetics of the T cell antigen receptor and T cell antigen recognition, *Ann. Rev. Immunol.* **4:**529–591.

Lanier, L.L., Cwirla, S., Felderspiel, N., and Phillips, J.H., 1986, Human natural killer cells isolated from peripheral blood do not rearrange T cell antigen receptor beta chain genes, *J. Exp. Med.* **163:**209–214.

LeFranc, M.P., Forster, A., Baer, R., Stinson, M.A., and Rabbitts, T.H., 1986a, Diversity and rearrangement of the human T cell rearranging gamma genes: nine germ line variable genes belonging to two subgroups, *Cell* **45:**237–246.

Lefranc, M.-P., Forster, A., and Rabbitts, T.H., 1986b, Rearrangement of two distinct T cell gamma chain variable region genes in human DNA, *Nature* **319:**420–422.

Lefranc, M., and Rabbitts, T.H., 1985, Two tandemly organized human genes encoding the T cell gamma constant region sequences show multiple rearrangement in different T cell types, *Nature* **316:**464–466.

Lewis, W.H., Michaelopoulos, E.E., Williams, D.L., Minden, M.D., and Mak, T.W., 1985, Two T cell leukemia patients with chromosomal translocations have breakpoints within the human T cell antigen receptor alpha chain locus, *Nature* **317**:544–546.

Linch, D.C., Berliner, N., O'Flynn, K., Kay, L.A., Jones, H.M., MacLennan, K., Heuhns, E.R., and Goff, K., 1985, Hodgkin cell leukemia of B cell origin, *Lancet* **i**:78–80.

Minden, M.D., Toyanaga, B., Ha, K., Yanagi, Y., Chin, B., Gelfand, E., Mak, T.W., 1985, Somatic rearrangement of T cell antigen receptor gene in human T cell malignancies, *Proc. Natl. Acad. Sci. USA* **82**:1224–1227.

Morton, C.C., Duby, A.D., Eddy, R.L., Shows, T.B., and Seidman, J.G., 1985, Genes for beta chain of human T cell antigen receptor map to regions of chromosomal rearrangement in T cells, *Science* **228**:582–585.

Murre, C., Waldmann, R.A., Morton, C.C., Bongiovanni, K.F., Waldmann, T.A., Shows, T.B., and Seidman, J.G., 1985, Human gamma chain genes are rearranged in leukemic T cells and mapped to the short arm of chromosome 7, *Nature* **316**:549–552.

O'Connor, N.T.J., Weatherall, D.J., Feller, A.C., Jones, D., Pallesen, G., Stein, H. Wainscoat, J.S., Gatter, K.C., Isaacson, P., Lennert, K., Ramsey, A., Wright, D.H., and Mason, D.Y., 1985, Rearrangement of the T cell receptor beta chain gene in the diagnosis of lymphoproliferative disorders, *Lancet* **i**:1295–1297.

Ohashi, P.S., Mak, T.W., Van den Elsen, P., Yanagi, Y., Yoshikai, Y., Calman, A.F., Terhorst, Stobo, J.D., and Weiss, A., 1985, Reconstitution of an active surface T3 T cell antigen receptor by DNA transfer, *Nature* **315**:606–609.

Pelicci, P.-G., Knowles, D.M., and Dalla-Favera, R., 1985, Lymphoid tumors displaying rearrangements of both immunoglobulin and T cell receptor genes, *J. Exp. Med.* **126**:1015–1024.

Prittaluga, S., Raffeld, M., Lipford, E.H., and Cossman, J., 1986, 3A1 (CD 7) expression precedes T beta gene rearrangements in precursor T (lymphoblastic) neoplasms, *Blood* **68**:134–439.

Quertermous, A., Murre, C., Dialynas, D.P., Duby, A.D., Strominger, J.L., Waldmann, T.A., and Seidman, J.G., 1986, Human T cell gamma genes: organization, diversity and rearrangement, *Science* **231**:252–255.

Rabbitts, T.H., Lefranc, M.P., Stinson, M.A., Sims, J.E., Schroeder, J., Steinmetz, M., Spurr, N.L., Solomon, E., and Goodfellow, P.N., 1985a, The chromosomal location of T cell receptor genes and a T cell rearranging gene: Possible correlation with specific translocations in human T cell leukemia, *EMBO J.* **4**:1461–1465.

Rabbitts, T.H., Stinson, A., Forster, A., Foroni, L., Luzzatto, L., Catovsky, D., Hammarstrom, L., Smith, C.I.E., Jones, D. Karpas, A. Minowada, J., and Taylor, A.M.R., 1985b, Heterogeneity of T cell beta chain gene rearrangements in human leukemias and lymphomas, *EMBO J.* **4**:2217–2224.

Rambaldi, A., Pelicci, P.-G., Allavena, A., Knowles, D.M., II, Rossini, S., Bassan, R., Barbui, T., Dalla-Favera, R., and Mantovani, A., 1985, T cell receptor beta chain gene rearrangements in lymphoproliferative disorders of large granular lymphocytes/natural killer cells, *J. Exp. Med.* **162**:2156–2162.

Raulet, D.H., Garman, R.D., Saito, H., and Tonegawa, S., 1985, Developmental regulation of T cell receptor gene expression, *Nature* **314**:103–107.

Rowley, J.D., 1978, The cytogenetics of acute leukemia, *Clin. Hematol.* **7**:385–406.

Sadamori, N., Kusano, M., Nishino, K., Tagawa, M., Yao, E.-I., Yamada, Y., Amagasaki, T., Kinoshita, K.-I., and Ichimaru, M., 1985, Abnormalities of chromosome 14 at band 14q11 in Japanese patients with adult T cell leukemia, *Cancer Genet.* **17**:279–282.

Sangster, R., Minowada, J., Suciu-Foca, N., Minden, M., and Mak, T.W., 1986, Rearrangement and expression of alpha, beta and gamma chain T cell receptor gene in human leukemic and functional T cell lines, *J. Exp. Med.* **163**:1491–1508.

Schwartz, D.C., and Cantor, C.R., 1984, Separation of yeast chromosome sized DNAs by pulsed field gradient gel electrophoresis, *Cell* **37**:67–75.

Shima, E.A., LeBeau, M.M., McKeithan, T.W., Minowada, J., Showe, L.W., Minden, M.D., Rowley, J.D., and Diaz, M.O., 1986, T cell receptor alpha chain gene moves immediately downstream of c-myc in a chromosomal 8;14 translocation in a cell line from a human T cell leukemia, *Proc. Natl. Acad. Sci. USA* **83**:3439–3443.

Sim, G.K., Yague, J., Nelson, J., Marrack, P., Palmer, E., Augustin, A., and Kappler, J., 1984, Primary structure of human T cell receptor alpha chain, *Nature* **132**:771–775.

Sims, J.E., Tunnacliffe, A., Smith, W.J., and Rabbitts, T.H., 1984, Complexity of human T cell antigen receptor beta chain constant and variable region genes, *Nature* **312**:541–545.

Siu, G., Clark, S., Yoshikai, Y., Malissen, M., Yanagi, Y., Strauss, E., Mak, T.W., and Hood, L., 1984, The human T cell antigen receptor is encoded by variable, diversity and joining gene segments that rearrange to generate a complete V gene, *Cell* **37**:393–401.

Sklar, J., Cleary, M.L., Thielemann, K., Gralow, J., Warnke, R., and Levy, R., 1984, Biclonal B cell lymphoma, *N. Eng. J. Med.* **311**:20–27.

Smith, L.J., Curtis, J.E., Messner, H.A., Senn, J.S., Furthmayr, H., and McCulloch, E.A., 1983, Lineage infidelity in acute leukemia, *Blood* **61**:1138–1145.

Smith, S.D., Morgan, R., Link, M.P., McFall, F., and Hecht, F., 1986, Cytogenetic and immunophenotypic analysis of cell lines established from patients with T cell leukemia/lymphoma, *Blood* **67**:650–656.

Snodgrass, R., Kisielow, P., Kiefer, M., Steinmetz, M., and von Boehmer, H., 1985, Ontogeny of the T cell antigen receptor within the thymus, *Nature* **313**:592–595.

Southern, E.M., 1975, Detection of specific sequences among DNA fragments separated by gel electrophoresis, *J. Mol. Biol.* **98**:503–517.

Tawa, A., Hozumi, N., Minden, M.D., Mak, T.W., and Gelfand, E.W., 1985, Rearrangement of the T cell receptor beta chain gene in non-T, non-B acute lymphoblastic leukemia of childhood, *N. Eng. J. Med.* **313**:1033–1037.

Tonegawa, S., 1983, Somatic generation of antibody diversity, *Nature* **302**:575–581.

Toyanaga, B., Yanagi, Y., Suciu-Foca, M., Minden, M., and Mak, T.W., 1984, Rearrangements of T cell receptor gene YT35 in human DNA from thymic leukemia T cell lines and functional T cell clones, *Nature* **311**:385–387.

Toyanaga, B., Yoshikai, Y., Vadasz, V., Chin, B., and Mak, T.W., 1985, Organization and sequences of a diversity, constant and joining region gene for human T cell receptor beta chain, *Proc. Natl. Acad. Sci. USA* **82**:8624–8628.

Waldmann, T.A., Davis, M.M., Bongiovanni, K.F., and Korsmeyer, S.J., 1985, Rearrangements of genes for the antigen receptor on T cells as markers of lineage and clonality in human lymphoid neoplasms, *N. Eng. J. Med.* **313**:776–783.

Weiss, A., and Stobo, J.D., 1983, Requirement for the coexpression of T3 and the T cell antigen receptor on a malignant human T cell line, *J. Exp. Med.* **160**:1284–1299.

Weiss, L.M., Strickler, J.G., Warnke, R.A., and Sklar, J., 1986, Immunoglobulin gene rearrangements in Hodgkin's disease, *Hum. Pathol.* **17**:1009–1014.

Weiss, L.M., Wood, G.S., Warnke, R.A., and Sklar, J., 1986, Clonal T cell populations in lymphomatoid papulosis: evidence of a lymphoproliferative origin for a clinically benign disease, *N. Eng. J. Med.* **315**:475–479.

Weiss, L.M., Hu, E., Wood, G.S., Moulds, C., Cleary, M.L. Warnke, R., and Sklar, J., 1985a, Clonal rearrangements of T cell receptor genes in mycosis fungoides and dermatopathic lymphadenopathy, *N. Eng. J. Med.* **313**:539–544.

Weiss, L.M., Trela, M.J., Cleary, M.L., Turner, R.R., Warnke, R.A., and Sklar, J., 1985b, Frequent immunoglobulin and T cell receptor gene rearrangements in histiocytic neoplasms, *Am. J. Pathol.* **121**:369–373.

Williams, D.L., Look, A.T., Melvin, S.L., Robertson, P.K., Dahl, G., Flake, T., and Stass, S., 1984, New chromosomal translocations correlate with specific immunophenotypes of childhood acute lymphoblastic leukemia, *Cell* **36**:101–109.

Yanagi, Y., Chan, A., Chin, B., Minden, M.D., and Mak, T.W., 1985, Analysis of cDNA clones specific for human T cells and the alpha and beta chains of the T cell receptor heterodimer from a human T cell line, *Proc. Natl. Acad. Sci. USA* **82:**3430–3434.

Yanagi, Y., Yoshikai, Y., Leggett, K., Clark, S.P., Aleksander, I., and Mak, T.W., 1984, A human T cell specific cDNA clone encodes a protein having extensive homology to immunoglobulin chains, *Nature* **309:**145–149.

Yancopoulos, G.D., Blackwell, T.K., Suh, H., Hood, L., and Alt., F.W., 1986, Introduced T cell receptor variable gene segments recombine in pre-B cells: evidence that B and T cells use a common recombinase, *Cell* **44:**251–259.

Yoshikai, Y., Anatoniou, A., Clark, S.P., Yanagi, Y., Sangster, R., Elsen, P., Terhorst, C., and Mak, T.W., 1984, Sequence and expression of two distinct human T cell receptor beta chain genes, *Nature* **315:**521–524.

Yoshikai, Y., Clark, S.P., Taylor, S., Sohn, U., Minden, M.D., and Mak, T.W., 1985, Organization and sequence of the variable, joining and constant region genes of the human T cell receptor alpha chain, *Nature* **316:**837–840.

Yoshikai, Y., Kimura, N., Toyanaga, B., and Mak, T.W., 1986, Sequences and repertoire of human T cell receptor alpha chain variable region genes in mature T lymphocytes, *J. Exp. Med.* **164:**90–103.

Zech, L., Gahrton, L., Hammarstrom, L., Juliusson, G., Mellstedt, H., Robert, K.H., and Smithe, C.I.E., 1984, Inversion of chromosome 14 marks human T cell chronic lymphocytic leukemia, *Nature* **308:**858–860.

Zehnbauer, B.A., Pardoll, D.M., Burke, P.J., Graham, M.L., and Vogelstein, B., 1986, Immunoglobulin gene rearrangements in remission bone marrow specimens from patients with acute lymphoblastic leukemia, *Blood* **67:**835–838.

9

The Involvement of the T-Cell Receptor in Chromosomal Aberrations

ILAN R. KIRSCH and GREGORY F. HOLLIS

1. The Involvement of the T-Cell Receptor in Chromosomal Aberrations

It is now clear that the T-cell receptor gene loci are directly involved in some of the most characteristic chromosomal aberrations specifically associated with T-cell disorders. This recent conclusion, however, should be viewed in the general context of the genesis of cell-type-specific chromosomal abnormalities. Over the past 25 years cytogeneticists have been describing an ever-increasing number of distinctive chromosomal abnormalities specifically associated with certain types of cancer (Yunis *et al.*, 1983). We are not speaking of the constitutional syndromes of karyotype abnormality (e.g., trisomy 21) found in every cell of an affected individual, which are believed to predispose the person to the development of certain cancers. Rather we are considering here those chromosomal abnormalities found only in the tumors of an otherwise karyotypically normal person. One of the earliest identified and best known associations is the occurrence of the reciprocal translocation between chromosomes 9 and 22 yielding as one of the derivative partners the so-called "Philadelphia" chromosome seen in over 95% of the tumors of patients with chronic myelogenous leukemia (Nowell and Hungerford, 1960; Rowley, 1980). Other different, but consistent chromosomal aberrations have now been reported for Burkitt's lymphoma (Zech

ILAN R. KIRSCH ● National Cancer Institute–Navy Medical Oncology Branch, National Cancer Institute, National Institutes of Health, Bethesda, Maryland 20892.   GREGORY F. HOLLIS ● Monsanto Co., St. Louis, Missouri 63198.

et al., 1976), a spectrum of additional hematopoietic malignancies (Chaganti, 1983), as well as a number of solid tumors (Yunis *et al.*, 1983). It is not clear that each and every one of these karyotypic abnormalities is causally related to the tumor in which it is found, but, given their distinctiveness and specificity, it is compelling that they represent some unique physiologic aspect of the cell types in which they occur.

2. The Burkitt's Lymphoma Model

Within the past five years our understanding of some of these associations of chromosomal aberrations with specific cancers has greatly increased as cytogenetic description has given way to molecular characterization. Currently, the best studied examples of specific chromosomal abnormalities associated with a disease state are those seen in Burkitt's lymphomas and mouse plasmacytomas. Burkitt's lymphoma is a tumor of immunoglobulin expressing B lymphocytes seen primarily in a pediatric age group (Ziegler, 1981). A brief review of the findings in Burkitt's lymphoma (Klein, 1983; Leder *et al.*, 1983) offers a framework for comparison when later we consider chromosomal aberrations that occur in T cells. In approximately 80% of Burkitt's tumors a distinctive reciprocal translocation occurs between chromosomes 8 and 14, t(8;14)(q24;q32) (Fig. 1). In the remaining 20% of these tumors "variant" translocations either between chromosomes 8 and 2, t(2;8)(p11;q24) or chromosomes 8 and 22, t(8;22) (q24;q11) are found. The genes that precisely flank the breakpoints of these translocations have now been identified and characterized on a molecular level (Shen-Ong *et al.*, 1982; Dalla-Favera *et al.*, 1982; Taub *et al.*, 1982). Translocations in Burkitt's lymphoma involve rearrangements of a region in which is encoded the c-*myc* oncogene, with one of the regions in which is encoded an immunoglobulin gene, kappa (2p11) (Malcolm *et al.*, 1982) heavy chain (14q32) (Kirsch *et al.*, 1982), or lambda (22q11) (McBride *et al.*, 1982; de la Chapelle *et al.*, 1983). In short, an oncogene c-*myc*, formerly discrete and on a separate chromosome is now brought into the context (and often made contiguous with) an immunoglobulin gene. This has occurred within a cell (B lymphocyte) whose primary differentiated function is the production of immunoglobulins. With this realization have come intense investigations exploring: (1) The relationship of these translocations to lymphomagenesis, and (2) the generalizability of the mechanism that produces this translocation to the development of other chromosomal aberrations.

2.1. Are the Translocations of Burkitt's Lymphoma Causal of the Tumor?

The c-*myc* protooncogene was implicated six years ago as being involved in lymphomagenesis in chickens (Hayward *et al.*, 1981). *myc* has been shown to

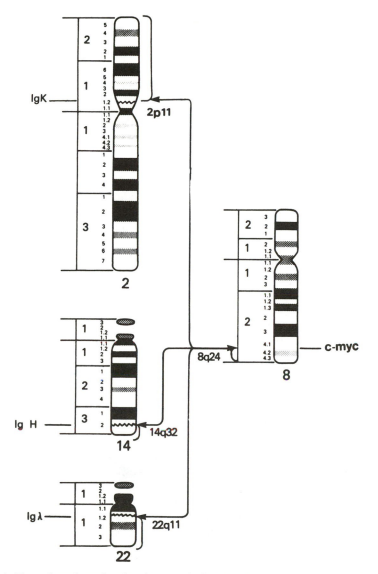

Figure 1. The reciprocal translocations between the Ig loci and c-*myc* protooncogene in Burkitt's lymphoma. Reproduced with permission of *Annals of Internal Medicine*. (From Kirsch, 1985b.)

function as one of the oncogenic partners in the "two-hit" or cooperative malignant transformation of rat embryo fibroblasts in culture (Land *et al.*, 1983). Recent experiments introducing a construct containing the human c-*myc* protooncogene into mouse embryos has resulted in subsequent tumor formation in function as one of the oncogenic partners in the "two-hit" or cooperative malignant transformation of rat embryo fibroblasts in culture (Land *et al.*, 1983). Recent experiments introducing a construct containing the human c-*myc* protooncogene into mouse embryos has resulted in subsequent tumor formation in tissues expressing this gene in the transgenic adults (Stewart *et al.*, 1984; Adams *et al.*, 1985; Leder *et al.*, 1986). The contribution of c-*myc* to tumor formation in these examples is believed to be mediated through inappropriate expression of its otherwise physiologic function.

There is now highly suggestive though still inferential evidence that the translocation of c-*myc* in Burkitt's lymphoma contributes to malignant transformation of the cell. C-*myc* is felt to be inappropriately expressed as a result of the Burkitt's lymphoma translocations. This speculation is derived from evidence indicating that in the tumor cells only the translocated c-*myc* is expressed, the gene on the uninvolved chromosome 8 is quiescent (Taub *et al.*, 1984). Furthermore there is an alteration in promoter utilization of the c-*myc* gene. Normally the c-*myc* protooncogene is transcribed from two distinct promoters, P1 and P2 with P2 utilization predominating over P1. In Burkitt's lymphoma promoter utilization shifts from P2 predominance and in some cases utilizes entirely distinct cryptic promoters (Taub *et al.*, 1984). Recent evidence has suggested the possibility that introgenic attenuation of c-*myc* transcripts may be relevant to control of this gene and altered in different cellular contexts (Nepveu and Marcu, 1986; Bentley and Groudine, 1986). Abrogation of this attenuation may be a mechanism of c-*myc* deregulation in some Burkitt's lymphomas (Dalla-Favera, personal communication.) There is some evidence that in these particular B-cell malignancies carrying the t(8;14), t(2;8), or t(8;22), the translocated c-*myc* is regulated similarly to an immunoglobulin gene (Showe and Croce, 1987).

2.2. What is the Mechanism of Translocation in Burkitt's Lymphoma and Other B-Lymphocyte Tumors?

In the more common (8;14) translocations c-*myc* is usually introduced into the immunoglobulin (Ig) heavy-chain locus (Ravetch *et al.*, 1981) by a recombination event within the heavy-chain "switch" regions or joining (J) regions (Dalla-Favera *et al.*, 1982; Taub *et al.*, 1982; Neri *et al.*, 1988). J regions are targets for the site-specific structural rearrangements that are a prerequisite to the formation of a functional immunoglobulin gene. The switch regions are the areas involved in physiologic switching of the humoral immune response from the IgM isotype to IgG, IgE, or IgA (Davis *et al.*, 1980; Kataoka *et al.*, 1980; Maki *et al.*, 1980; Ravetch *et al.*, 1980). Switch regions are found proximal to each Ig heavy-chain subclass and are believed to interact via an enzyme-mediated pseudo-homologous recombination event (Shimizu and Honjo, 1984). Because the function of the J and switch regions in lymphocytes appears to involve their participation in DNA breakage and rejoining events within the immunoglobulin heavy-chain locus, it is tempting to assume that the majority of translocations in Burkitt's lymphomas result from these same recombination systems going awry and substituting a *myc* gene instead of a variable region or different Ig subtype as a coparticipant in the recombination event. Because these recombinations occur at very specific times in B-cell development, these results may define the developmental time frame in which Burkitt's lymphoma cells are generated. This view however must be tempered by the knowledge that no definitive J signal sequences, nor switch region homologous areas have been found within the germline c-*myc* gene locus (Battey *et al.*, 1983). The mechanism of breakage and rejoining is less clear in the variant translocations. Detailed molecular analysis of the points of breakage and rejoining in these variant translocations may not support involvement of the J or switch recombination systems in the occurrence of the rearrangements (Hollis *et al.*, 1984; Denny *et al.*, 1985).

Burkitt's lymphoma likely represents a tumor of B lymphocytes advanced beyond the pre-B-cell stage in their developmental pathway. Lymphomas of B cells displaying genotypic and phenotypic evidence of earlier developmental stage also carry distinctive chromosomal abnormalities. Of this group two examples also suggest a mechanism for the occurrence of specific translocation. The t(14;18)(q32;q21) translocation seen in follicular lymphomas and the t(11;14)(q24;q32) seen in diffuse small and large cell lymphomas both show evidence of involvement of an immunoglobulin recombination system in their formation (Tsujimoto *et al.*, 1984a; Tsujimoto *et al.*, 1984b; Tsujimoto *et al.*, 1985; Bakhshi *et al.*, 1985). In these examples the breakpoint on chromosome 14 occurs in close proximity to the heavy-chain joining (J) sequence. It therefore seems highly likely that the same system that mediates the physiologic variable (V)-J site-specific recombination event that is a prerequisite to the formation of a functional immunoglobulin (Brack *et al.*, 1978; Seidman and Leder, 1978) is playing a role in the aberrant recombination events that result in these translocations. Again, as in the example of Burkitt's lymphoma, this conclusion must be viewed with some caution. Although the recombination event does often occur at an appropriate target site *vis á vis* the J sequences, recombination signal sequences analogous to those surrounding V regions and believed crucial to the recombination event are not always found with a high degree of fidelity within the involved regions of chromosome 11 or 18 (Tsujimoto *et al.*, 1985; Bakhshi *et al.*, 1985).

In summary, an extant enzyme system with a developed DNA rearranging capacity may participate in mediating translocation events in cells at a stage in their development when the system is normally active, but such systems appear neither necessary nor sufficient to explain the diverse associations of chromosomal abnormalities and cancers in general. Still, there is an important concept inherent in the observation that the expression of certain genes during cellular development affects the mechanism of chromosomal rearrangements in these cells. This can be related either to products made in the cell that effect the translocation event, or the targets of those effectors. In hematopoietic cells the target regions of chromosomal aberrations can often be shown to encode genes active in the differentiated functions carried out by the cells (Kirsch *et al.*, 1985a). In B lymphocytes those genes would include the immunoglobulins, in erythroid precursors, globins, and in T lymphocytes, T-cell receptors. We will make reference to this concept again when we discuss the molecular characterization of chromosomal aberrations that involve the T-cell receptor gene encoding regions.

3. The Karyotypic Abnormalities Characteristic of T-Cell Disorders

Studies more than eight years old are difficult to assess given the more primitive cytogenetic techniques used for the analyses. It has only been within this recent time frame that technical improvements in cell synchronization, spreading of early metaphase chromosomes and staining techniques have achieved a degree of sophistication so that high resolution analyses of human chromosomes are now possible [Yunis and Chandler (Harper), 1977]. Even so, complex karyotypic abnormalities can still defy identification. It is with this perspective that the recent literature can be reviewed in search of common, consistent chromosomal abnormalities seen in T cells and T-cell disorders.

A given karyotypic analysis on one T-cell malignancy may reveal numerous and diverse chromosomal abnormalities quite distinct from those revealed in an analysis of a different patient's tumor. In a summation of studies of 105 patients with malignant T-cell diseases, Ueshima *et al.* (1984) found that every chromosome had, in one patient or another, been reported as being gained or lost. In this same summation it was found that every chromosome had, in one patient or another, been involved in a structural rearrangement. However, structural alterations of chromosome 14 predominated in this population being seen in 40 of the 105 patients from the literature review. In a subgroup of those patients on whom more information was available, 54% had a breakpoint in the upper segment of the long arm.

In more recent studies the predominance of two segments, 14q11-13, and 14q32 as breakpoints in chromosomal aberrations seen in T-cell malignancies is confirmed. Thus, of 16 patients with E-rosette positive T-cell ALL reported by Williams *et al.* (1984), four of the tumors carried the reciprocal translocation t(11;14)(p13;q13) and one carried the t(8;14)(q24;q12). Zech *et al.* (1984) found a characteristic inversion of chromosome 14, inv(14)(q11;q32) in four of five patients with T-cell CLL. Recent improved classification techniques suggest that this inversion may be particularly associated with a more virulent form of T-cell CLL representing cells transformed at the prothymocyte stage (Brito-Babapulle *et al.*, 1987; Gale and Rai, 1987). Hecht *et al.* (1984) observed this identical chromosomal anomaly in another patient with T-cell CLL as well as in a cell line derived from a pediatric patient with T-cell lymphoma. A cell line derived from a different patient with T-cell lymphoma showed involvement of chromosome 14 in a reciprocal translocation with chromosome 10, t(10;14)(q23;q11.2) (Smith *et al.*, 1986). Sadamori *et al.* (1985) studied 11 patients with "typical" adult T-cell leukemia and found involvement of chromosome 14 (band q11-13) in ten. Of these, five were translocations between two chromosomes 14, t(14;14)(q11-13;q32). One was the inv(14)(q11;q32) mentioned above. Also seen were a t(11;14) (p11;q11), a t(12;14)(q24;q11), as well as two other only partially characterized abnormalities both of which however could be shown to involve band 14q11-13. Hershfield *et al.* (1984) have reported a patient with an initial T-cell leukemia whose clinical picture merged into that of a promyeloctic leukemia apparently in response to therapy with deoxycoformicin. This patient's tumor cells and a cell line derived from the tumor carried a translocation, t(1;14)(p33;q11) (Kurtzberg *et al.*, 1985). A t(14;14)(q12;q31) has been observed in malignant cells from a patient with Sezary syndrome (Shah-Reddy *et al.*, 1982). Miyamoto *et al.* (1983) reported karyotypic abnormalities involving the more telomeric position on chromosome 14, 14q32, in eight of 15 patients with adult T-cell leukemia. Partial deletions of the long arm of chromosome 6, 6q, were noted in seven patients in this same study. This deletion of 6 has also been reported by Whang-Peng (1985) in a study of patients with HTLV-positive lymphomas. An earlier study of a "healthy" adult with positive serology for adult T-cell leukemia-associated antigens (Fukuhara *et al.*, 1983) revealed a clonal abnormal T-cell population carrying a t(7;14)(p15;q13).

This last case introduces another chromosome, chromosome 7, frequently involved in karyotypic abnormalities seen in T-cell disorders. A t(7;9) (q34;q33-34) was reported in three T-cell leukemia and lymphoma cell lines (Hecht *et al.*, 1985). Trisomy of chromosome 7 (Ueshima *et al.*, 1981) has been suggested to be a recurring finding in certain subtypes of T-cell ALL. Abnormalities involving two bands 7p13-15 and 7q32-35, on chromosome 7 and the previously described regions of chromosome 14 have long been observed in T cells from "normal" persons (Welch and Lee, 1975; Beatty-DeSana *et al.*, 1975; Hecht *et al.*, 1975) and particularly in T cells from patients with the disease ataxia-telangiectasia (A-T)(Aurias *et al.*, 1980; Scheres *et al.*, 1980; Wake *et al.*, 1982). A-T is a disease of protean manifestations (McFarlin *et al.*, 1972) including a tendency to chromosomal breaks. The abnormalities often involve translocations or inversions (Taylor, 1982) with breakpoints within these bands. Often chromosomes 7 and 14 participate as reciprocal partners in the translocations. Again as in the malignant disorders discussed above, abnormalities of chromosome 14 with breakpoints within the 14q11-12 region are the most consistent finding in T cells from patients with A-T (Kaiser-McCaw *et al.*, 1975; Kohn *et al.*, 1982; Taylor, 1982). It is now beginning to appear that the 14q32 breakpoints seen in the inv(14) abnormality in normal stimulated T cells and in some but not all inv(14) tumors differs from the 14q32 breakpoint seen as a proliferative but not frankly malignant outgrowth of T cells in the peripheral blood of patients with ataxia-telangiectasia (Kennaugh *et al.*, 1986; Mengele-Gaw *et al.*, 1987; A. Aurias and M. Stern, personal communication, A.M.R. Taylor, personal communication). Such afflicted patients may develop a karyotypic, immunophenotypic, and genotypic T-cell clone in their peripheral blood that can comprise greater than 90% of the phytohemagglutinin responsive population. Remarkably, and of possible importance from the standpoint of T-cell function and compartmentalization, such patients can survive for years without developing either lymphoid malignancy or the spectrum of opportunistic infections characteristics of other T-cell deficiency states (A. Aurias, R. Gatti, F. Hecht, I. Kirsch, M.H. Stern, A.M.R. Taylor, unpublished data).

4. The Chromosomal Locations of the T-Cell Receptor Alpha, Beta, and Gamma Chains—Relevance to Chromosomal Aberrations That Highlight Regions of Genomic Activity in Differentiated Cells

Localizations of the immunoglobulin gene loci to those regions frequently involved in the chromosomal abnormalities seen in B-cell tumors provided a provocative clue to deciphering the molecular events that had occurred in these tumors. It led to the conclusion that the primary differentiated product of B cells, immunoglobulins, were intimately involved in the translocation events by which tumors of this cell type were distinguished. The functional analogy between immunoglobulin and T-cell receptor genes has been assumed for a decade and certainly strengthened recently by the obvious structural similarity of these loci. Thus it was with great anticipation that the genomic localizations of the T-cell receptor loci were sought, with particular attention being focused on the relationship of these localizations to the chromosomal aberrations commonly seen in T-cell disorders (see Section 3). The prediction of many investigators has now been fully born out. The three T-cell receptor loci map to three of the four most common breakpoints seen in chromosomal aberrations of T cells (Fig. 2). The α chain of the human T-cell receptor has been mapped to chromosome 14 (Collins *et al.*, 1985; Jones *et al.*, 1985) band 14q11-12 (Caccia *et al.*, 1985; Croce *et al.*, 1985). The human β-chain locus resides on chromosome 7 within the region 7q32-35 (Isobe *et al.*, 1985; LeBeau *et al.*, 1985; Morton *et al.*, 1985). The human γ chain has also been localized to chromosome 7, band 7p14-15 (Murre *et al.*, 1985). Cloning data support the localization of the newly described δ chain to band 14q11.2. It is obvious that there is some ambiguity about the precise bands involved in the karyotypic abnormalities as well as the precise bands of localization of the T-cell receptor loci. This probably reflects the limit of specificity of current cytogenetic analyses in the hands of the various investigators carrying out these studies. It is very possible that the ambiguity of

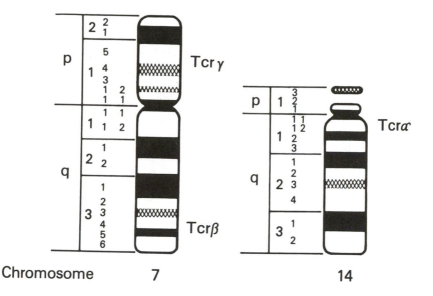

Figure 2. The chromosomal localizations of the T-cell receptor α-, β-, and γ-chain genes. The δ-chain gene, not shown in this figure, also is located at band 14q11.2.

band localization and the slight broadening of these assignments are technical issues and do not represent an actual distribution of breakpoints or gene loci across the whole of the chromosome segments that have been delimited. Molecular analyses have now demonstrated the direct involvement of the α and β T-cell receptor gene loci in chromosomal breaks in T cells (see Section 5). It is certainly possible that every chromosomal abnormality involving these regions will turn out to fall within the relevant T-cell receptor locus and that the scatter of breakpoints currently reported within these regions may be artifactual.

5. Cloning Studies on Chromosomal Aberrations in T Cells

Recently, two separate research groups have taken a T-cell line described by Hecht *et al.* (1984) and demonstrated direct involvement of the T-cell receptor α-chain locus in a chromosomal aberration (Baer *et al.*, 1985; Denny *et al.*, 1986a). The line was derived from a pleural effusion of a 6-year-old male with lymphoblastic lymphoma and was shown to carry a distinctive associated chromosomal abnormality consisting of an inversion of chromosome 14 with breakpoints within the region where the T-cell receptor α chain had been localized, 14q11.2, and within the region where the immunoglobulin heavy-chain locus resides, 14q32 (Kirsch *et al.*, 1982). Southern blot analyses using probes that included distinctly identified T-cell receptor α-chain joining (J) segments

(Yoshikai *et al.*, 1985) revealed three separate J-chain rearrangements. Each of these were cloned. Two rearrangements had been caused by site specific recombination with variable (V) segments from the T-cell receptor α-chain locus. The genomic DNA 3' to one of these rearrangements had been disrupted so that the $V_\alpha J_\alpha$ region was no longer linked to the alpha constant region. The other $V_\alpha J_\alpha$ region was linked approximately 15 kb upstream of an α constant region but the $V_\alpha J_\alpha$ recombination event had occurred such that a frame-shift took place at the V-J join (Fig. 3A).

The cloning of the third α-chain J rearrangement provided the key to understanding the inversion event. This J had also undergone site-specific recombination but not with any element from the T-cell receptor α-chain loci. Instead an immunoglobulin heavy-chain variable region had been brought inframe into the center of the T-cell receptor locus (Fig. 3B). Knowing the orientation of these two loci on chromosome 14 (Erikson *et al.*, 1982; Erikson *et al.*, 1985) a model for the occurrence of the chromosomal inversion became obvious. Site-specific recombination between an immunoglobulin variable region and a T-cell receptor J segment would result in a chromosomal breakage and rejoining event that would precisely cause the morphologic alteration of chromosome 14 that had been described (Fig. 4). Because of the orientation of the immunoglobulin and T-cell receptor loci this model is unlikely to account for the occurrence of the related but distinct translocation between two chromosomes 14, t(14;14) (q12;q32), also associated with T-cell disorders. In the t(14;14) there is evidence for involvement of the α-chain T-cell receptor locus as one site of chromosomal breakage but the characterization of the site on 14q32 is incomplete at present although junction sequences do not appear to be derived from the immunoglobulin heavy chain locus (Mengele-Gaw *et al.*, 1987; A.M.R. Taylor, C.M. Croce, personal communication). The same situation exists for the analysis of a t(11;14)(p13;q11) in which direct involvement of the T-cell receptor α chain is highly likely (Erikson *et al.*, 1985; Lewis *et al.*, 1985).

Involvement of the T-cell receptor α-chain locus. is also suggested by an analysis of a line derived from a pediatric patient with T cell acute lymphoblastic leukemia carrying a translocation, t(8;14)(q24;q11) (Mathieu-Mahul *et al.*, 1985). The breakpoint on chromosome 8 occurred 3' of and in close proximity to the c-*myc* protooncogene. The cloned breakpoint on chromosome 14 did not show crosshybridization with a cDNA for the alpha chain of the T-cell receptor, but was seen to be presumatively rearranged in some other DNAs derived from patients with T-cell malignancies. It is possible that this segment might involve one of the many upstream J regions dispersed over greater than 50 kb in the α-chain locus. An analysis of two additional T-cell leukemias carrying this same translocation, t(8;14)(q24;q11), confirmed that the breakpoints occurred 3' of the c-*myc* oncogene and split the T-cell receptor (TcR) α-chain locus between the TcR variable and constant genes. Furthermore, this study suggests that the c-*myc* oncogene is deregulated by the translocation. As in the Burkitt's lym-

A

CT GTG GAA GGA GCC ATT GTC CAG ATA AAC TGC ACG TAC CAG ACA TCT GGG TTT TAT GGG CTG TCC TGG TAC CAG
   Val Glu Gly Ala Ile Val Gln Ile Asn Cys Thr Tyr Gln Thr Ser Gly Phe Tyr Gly Leu Ser Trp Tyr Gln

CAA CAT GAT GGC GGA GCA CCC ACA TTT CTT TCT TAC AAT GCT CTG GAT GGT TTG GAG GAG ACA GGT CGT TTT TCT
Gln His Asp Gly Gly Ala Pro Thr Phe Leu Ser Tyr Asn Ala Leu Asp Gly Leu Glu Glu Thr Gly Arg Phe Ser

TCA TTC CTT AGT CGC TCT GAT AGT TAT GGT TAC CTC CTT CTA CAG GAG CTC CAG ATG AAA GAC TCT GCC TCT TAC
Ser Phe Leu Ser Arg Ser Asp Ser Tyr Gly Tyr Leu Leu Leu Gln Glu Leu Gln MET Lys Asp Ser Ala Ser Tyr

TTC TGC GCT GTG CCC CCCT | GAT GGA TAG CAG CTA TAA ATT GAT CTT CGG GAG TGG GAC CAG ACT GCT GGT CAG
Phe Cys Ala Val Pro Pro | MET Asp Ser Ser Tyr Lys Leu Ile Phe Gly Ser Gly Thr Arg Leu Leu Val Arg

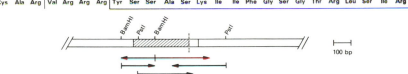

B

                                                                                    FR1
TC ACA GGG GTC CTG TCC CAG GTG CAG CTG CAG GAG TCG GGA CCA GGA CTG GTG AAG CCT TCG GAG ACC CTG TCC
   Val Leu Ser Gln Val Gln Leu Gln Glu Ser Gly Pro Gly Leu Val Lys Pro Ser Glu Thr Leu Ser

                                          CDR1   FR2
CTC ACC TGC ACT GTC TCT GGT TAC TCC ATC AGC | AGT GGT TAC TAC TGG GGC | TGG ATC CGG CAG CCC CCA GGG AAG
Leu Thr Cys Thr Val Ser Gly Tyr Ser Ile Ser | Ser Gly Tyr Tyr Trp Gly | Trp Ile Arg Gln Pro Pro Gly Lys

                                              CDR2
GGG CTG GAG TGG ATT GGG | AGT ATC TAT CAT AGT GGG AGC ACC TAC TAC AAC CCG TCC CTC AAG AGT | CGA GTC ACC
Gly Leu Glu Trp Ile Gly | Ser Ile Tyr His Ser Gly Ser Thr Tyr Tyr Asn Pro Ser Leu Lys Ser | Arg Val Thr

                              FR3
ATA TCA GTA GAC ACG TCC AAG AAC CAG TTC TCC CTG AAG CTG AGC TCT GTG ACC GCC GCA GAC ACG GCC GTG TAT
Ile Ser Val Asp Thr Ser Lys Asn Gln Phe Ser Leu Lys Leu Ser Ser Val Thr Ala Ala Asp Thr Ala Val Tyr

TAC TGT GCG AGA | GTC CGT CGG AGG | TAC AGC AGT GCT TCC AAG ATA ATC TTT GGA TCA GGG ACC AGA CTC AGC ATC CGG
Tyr Cys Ala Arg | Val Arg Arg Arg | Tyr Ser Ser Ala Ser Lys Ile Ile Phe Gly Ser Gly Thr Arg Leu Ser Ile Arg

Figure 3. (A) Nucleotide sequence and amino acid translation of the coding sequence of a Tcr V_α-J_α rearrangement in a T-cell lymphoma cell line. (B) Nucleotide sequence and amino acid translation of the coding sequence of an IgV_H-$TcrJ_\alpha$ rearrangement in a T-cell lymphoma cell line. Reproduced from *Nature*. (For details of this analysis see Baer *et al.*, 1985; Denny *et al.*, 1986a.)

phoma examples, only the c-*myc* on the translocated chromosome is expressed (Erikson *et al.*, 1986). This group has gone on to clone and characterize the breakpoint of a t(8;14)(q24;q11) from a leukemic cell line. A new junction occurs 3 kb 3′ of the c-*myc* protooncogene and at a TcR joining segment 36 kb 5′ of the constant region (Finger *et al.*, 1986).

Involvement of the c-*myc* protooncogene and α chain of the T-cell receptor in a translocation, t(6;14), is being investigated (C.B. Thompson, personal communication). Work in progress on a T-cell lymphoma line carrying a t(7;9)(q32;q34) has revealed direct involvement of the β-chain locus of the T-cell receptor with the precise position of the breakpoint on chromosome 9 still

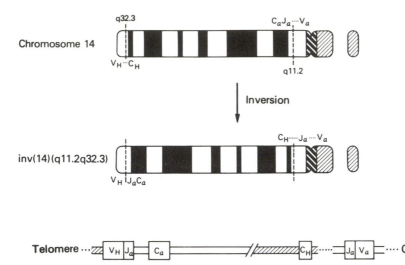

Figure 4. A model for inversion of chromosome 14 via IgV$_H$-TcrJ$_\alpha$ site-specific recombination. Reproduced from *Nature* (Denny *et al.*, 1986a.)

undetermined (J. Sklar, personal communication). Westbrook *et al.* (1987) have analyzed an identical morphologic inversion in a cell line derived from a patient with T-cell ALL. They found no DNA rearrangement within 250 kb of v-*abl* homologous sequences.

6. The Relevance of the Involvement of the T-Cell Receptor in Chromosomal Aberrations of T Cells

Investigations of the involvement of the T-cell receptor gene loci have so far yielded important information on chromosomal instability, lymphocyte development, and carcinogenesis. Just as in the Burkitt's lymphoma model with which this chapter began, one finds a primary differentiated product of a particular cell type (T-cell receptors in T cells) involved in characteristic chromosomal breakage and rejoining events. Again there is evidence that an extant recombination system active in the cell may be mediating at least part of the breakage and rejoining event.

The molecular analyses of the chromosome 14 inversion in the T-cell lymphoma line (Baer *et al.*, 1985; Denny *et al.*, 1986a) provide an important perspective on lymphocyte development. From these studies it is clear that a single V-J recombination system can recognize both the immunoglobulin and T-cell receptor loci and actually unite them into a hybrid gene. As a corollary to this observation it is apparent that at some stage of lymphocyte development both of these distinct and disparate gene systems, T-cell receptor α chain and immunoglobulin heavy chain, were simultaneously available and accessible to the site-specific recombination system. Recent molecular analysis of an identical morphologic inversion in the tumor cells of a patient with acute lymphocytic leukemia of B-cell lineage has revealed an identical mechanism of chromosomal recombination (Denny *et al.*, 1986b).

The cause and effect relationship of the T-cell receptor implicated chromosomal aberrations to the development of T cell malignancies is still unclear. It is possible that, as suggested by analogous B-cell studies, protooncogenes in a discrete number of chromosomal locations become activated or deregulated when brought into the new context of the T-cell receptor gene loci. This could certainly be the case in the t(8;14)(q24;q11) examples described by Mathieu-Mahul *et al.* (1985), Erikson *et al.* (1986), and Finger *et al.* (1986). The inv(14) example again must be looked to as well in this matter since it is one of the best characterized of the chromosomal aberrations associated with T-cell malignancies. As mentioned previously there are likely to be more than one type of inv(14) seen in T cells, some involving the TcR chain with an as yet uncharacterized gene on 14q32 (Mengele-Gaw *et al.*, 1987), others as described above, yielding a TcR α-IgH hybrid gene. This hybrid inversion may, of course, be incidental to the development of the lymphoma and only reflect the effect of genomic activity on chromosomal aberration. Since it is estimated that 0.1% of phytohemagglutinin-responsive cells in the peripheral blood of normal individuals carry such an inversion (Aurias *et al.*, 1985), it is possible to imagine that, fortuitously, such a cell will on occasion undergo malignant transformation. It might be that other breakpoints at 14q32 in T cells would occur near the same Ig heavy-chain V region whose sequence has been presented here. Alternatively, the hybrid gene might itself have oncogenic potential. The rearrangement occurs in frame and the hybrid gene is transcribed into poly A+ mRNA (Denny *et al.*, 1986a,b). Immunoglobulins and T-cell receptors are cell-surface receptors capable of transducing mitogenic stimuli at the cell surface into proliferative activity of the cells on which they are anchored. A hybrid gene such as this, capable of translation into a hybrid protein, might somehow convey an aberrant signal for proliferation to the cell. It is possible that in one or both of these types of inv(14) an as yet undescribed protooncogene resides near or is brought close to the particular immunoglobulin or TcR genomic regions involved in this inversion. A breakpoint region that might contain such a putative oncogene and that is reported to exist 3' of the IgH locus has been dubbed by some investigators "tc1-1" (C. Croce, personal communication). Such speculation will be directly tested as research continues on specific chromosomal aberrations in normal and malignant lymphoid cells.

7. Summary

The involvement of the T-cell receptor gene loci in chromosomal aberration occurring in T cells is unequivocal. Completely consistent association between

Table I. Translocations Involving T-Cell Receptor Genes

Disease	Translocation	Loci involved	Reference
Pre T(T-cell) leukemia/lymphoma	t(1;14)(p32;q11)	L-myc?-TcR$_\alpha$ N-ras?TcR$_\alpha$ src-2?-TcR$_\alpha$	Kurtzberg et al. (1985)
			C. Thompson, personal communication
	t(6;14)(q21;q11)	c-myb-TcR$_\alpha$	
	t(8;14)(q24;q11)	c-myc-TcR$_\alpha$	Williams et al. (1984) Mathieu-Mahul et al. (1985) Erikson et al. (198x) Finger et al. (1986)
	t(10;14)(q23;q11.2)	TdT?-TcR$_\alpha$ onc?TcR$_\alpha$	Smith et al. (1985)
	t(11;14)(p13;q11)	Wilm's?-TcR$_\alpha$ onc?TcR$_\alpha$	Williams et al. (1984) Sadamori et al. (1985) Erikson et al. (19xx) Lewis et al. (1985)
	t(12;14)(q24;q11)	onc?-TcR$_\alpha$	Sadamori et al. (1985)
	inv(14)(q11;q32)		Zech et al. (1984) Brito-Babapulle et al. (1987) Baer et al. (1985)

Disease	Aberration	Gene(s)	References
	t(14;14)(q11;q32)	IgH-TcR$_\alpha$ onc?-TcR$_\alpha$	Denny et al. (1986a) Mengele-Gaw et al. (1987)
		IgH?-TcR$_\alpha$ akt?TcR$_\alpha$ onc?-TcR$_\alpha$	Sadamori et al. (1985) Shah-Reddy et al. (1982) Mengele-Gaw et al. (1987)
	t(7;9)(q34;q33)	TcR$_\beta$-abl?	Hecht et al. (1985) J. Sklar, personal communication C. Westbrook et al. (1987)
Pre B(B-cell) leukemia/lymphoma	inv(14)(q11;32)	IgH-TcR$_\alpha$	Denny et al. (1986b)
Ataxia-telangiectasia or "stimulated" T cells	inv(7)(p15;q33) t(7;14)(p13;q11) t(7;14)(q34;q11) inv(14)(q11;q32) t(14;14)(q11;q32)	TcR$_\gamma$?-TcR$_\beta$ TcR$_\gamma$?-TcR$_\alpha$ TcR$_\beta$?-TcR$_\alpha$ TcR$_\alpha$?-IgH? TcR$_\alpha$?-IgH? TcR$_\alpha$ — a gene proximal to the IgH locus on 14q32	Wake et al. (1982) Fukuhara et al. (1983) Hecht et al. (1975) Aurias et al. (1980) Taylor (1982) Welch and Lee (1975) Beatty de Sana et al. (1975) Scheres et al. (1980) Kaiser-McCaw et al. (1975) Kennaugh et al. (1986)
	t(X;14)(q28;q11.2)	?-TcR$_\alpha$	Aurias and Stern (pers. commun.) Taylor (1982)

one of these loci and a karyotypic abnormality in a particular tumor (such as is seen for the immunoglobulin loci in Burkitt's lymphoma) is emerging as problems in tumor classification, cytogenetic analysis, and structural definition of these loci are being resolved. Studies of the T-cell receptor organization and function in relation to the development of these chromosomal aberrations is providing insights not only into questions of immunological import, but into basic issues of developmental and cancer biology.

References

Aurias, A., Dutrillaux, B., Buriot, D., and Lejeune, J., 1980, High frequencies of inversions and translocations of chromosomes 7 and 14 in ataxia telangiectasia, *Mutat. Res.* **69**:369–374.

Aurias, A., Couturier, J., Dutrillaux, A.-M., Dutrillaux, B., Herpin, F., Lamoliatte, E., Lombard, M., Muleris, M., Paravatou, M., Prieur, M., Prod'homme, M., Sportes, M., Viegas-Pequignot, E., and Volobouev, V. (1985). Inversion 14(q12q ter) or (q11.2q32.3): The most frequently acquired rearrangement in lymphocytes. *Hum. Genet.* **71**:19–21.

Adams, J.M., Harris, A.W., Pinkert, C.A., Corcoran, L.M., Alexander, W.S., Cory, S., Palmiter, R.D., and Brinster, R.L., 1985, The *c-myc* oncogene driven by immunoglobulin enhancers induces lymphoid malignancy in transgenic mice, *Nature* **318**:533–538.

Baer, R., Chen, K.-C., Smith, S.D., and Rabbitts, T.H., 1985, Fusion of an immunoglobulin variable gene and a T cell receptor constant gene in the chromosome 14 inversion associated with T cell tumors, *Cell* **43**:705–713.

Bakhshi, A., Jensen, J.P., Goldman, P., Wright, J.J., McBride, O.W., Epstein, A.L., and Korsmeyer, S.J., 1985, Cloning the chromosomal breakpoint of t(14;18) human lymphomas: Clustering around J_H on chromosome 14 and near a transcriptional unit on 18, *Cell* **41**:899–906.

Battey, J., Moulding, C., Taub, R., Murphy, W., Stewart, T., Potter, H., Lenoir, G., and Leder, P., 1983, The human *c-myc* oncogene: Structural consequences of translocation into the IgH locus in Burkitt lymphoma, *Cell* **34**:779–787.

Beatty-DeSana, J.W., Hoggard, M.J., and Cooledge, J.W., 1975, Non-random occurrence of 7-14 translocations in human lymphocyte cultures, *Nature* **255**:243–244.

Bentley, D.L., Groudine, M., 1986, A block to elongation is largely responsible for decreased transcription of *c-myc* in differentiated HL-60 cells, *Nature* **321**:702–706.

Brack, C., Hirama, M., Lenhard-Schuller, R., and Tonegawa, S., 1978, A complete immunoglobulin gene is created by somatic recombination, *Cell* **15**:1–14.

Brito-Babapulle, V., Pomfret, M., Matutes, E., and Catovsky, D., 1987, Cytogenetic studies on prolymphocytic leukemia. II. T cell prolymphocytic leukemia, *Blood* **70**:926–931.

Caccia, N., Bruns, G.A.P., Kirsch, I.R., Hollis, G.F., Bertness, V., and Mak, T.W., 1985, T cell receptor α chain genes are located on chromosome 14 at 14q11-14q12 in humans, *J. Exp. Med.* **161**:1255–1260.

Chaganti, R.S.K., 1983, Significance of chromosome change to hematopoietic neoplasms, *Blood* **62**:515–524.

Collins, M.K.L., Goodfellow, P.N., Spurr, N.K., Soloman, E., Tanigawa, G., Tonegawa, S., and Owen, M.J., 1985, The human T-cell receptor α-chain gene maps to chromosome 14, *Nature* **314**:273–274.

Croce, C.M., Isobe, M., Palumbo, A., Puck, J., Ming, J., Tweardy, D., Erikson, J., Davis, M., and Rovera, G., 1985, Gene of α-chain of human T-cell receptor: Location on chromosome 14 region involved in T-cell neoplasms, *Science* **227**:104–1047.

Dalla-Favera, R., Bregni, M., Erikson, J., Pattereson, D., Gallo, R.C., and Croce, C.M., 1982, Human *c-myc* oncogene is located on the region of chromosome 8 that is translocated in Burkitt lymphoma cells, *Proc. Natl. Acad. Sci. USA* **79**:7824–7827.

Davis, M., Calame, K., Early, P., Livant, D., Joho, R., Weissman, I., and Hood, L., 1980, An immunoglobulin heavy chain gene is formed by at least two recombinant events, *Nature* **283**:733–739.

de la Chapelle, A., Lenoir, G., Boue, J., Boue, A., Gallano, P., Huerre, C., Szajnert, M.-F., Jeanpierre, M., Lalonel, J.-M., and Kaplan, J.-C., 1983, Lambda Ig constant region genes are translocated to chromosome 8 in Burkitt's lymphoma t(8;22), *Nuc. Acids Res.* **11**:1133–1142.

Denny, C.T., Hollis, G.F., Magrath, I.T., and Kirsch, I.R., 1985, Burkitt lymphoma cell line carrying a variant translocation creates new DNA at the breakpoint and violates the hierarchy of immunoglobulin gene rearrangement, *Mol. Cell. Biol.* **5**:3199–3207.

Denny, C.T., Yoshikai, Y., Mak, T.W., Smith, S.D., Hollis, G.F., and Kirsch, I.R., 1986a, A chromosome 14 inversion in a T cell lymphoma is caused by site-specific recombination between immunoglobulin and T cell receptor loci, *Nature* **320**:549–551.

Denny, C.T., Hollis, G.F., Hecht, F., Morgan, R., Link, M.P., Smith, S.D., and Kirsch, I.R., 1986b, Common mechanism of chromosomal inversion in B and T cell tumors: Relevance to lymphocyte development, *Science* **234**:197–200.

Erikson, J., Finan, J., Nowell, P.C., and Croce, C.M., 1982, Translocation of immunoglobulin V_H genes in Burkitt lymphoma, *Proc. Natl. Acad. Sci. USA* **79**:5611–5615.

Erikson, J., Williams, D.L., Finan, J., Nowell, P.C., and Croce, C.M., 1985, Locus of the α-chain of the T cell receptor is split by chromosome translocation in T cell leukemias, *Science* **229**:784–786.

Erikson, J., Finger, L., Sun, L., ar-Rushdi, A., Nishikura, K., Minowada, J., Finan, J., Emanuel, B.S., Nowell, P.C., and Croce, C.M., 1986, Deregulation of *c-myc* by translocation of the α-locus of the T cell receptor in T cell leukemias, *Science* (in press).

Finger, L.R., Harvey, R.C., Moore, R.C.A., Showe, L.C., Croce, C.M., 1986, A common mechanism of chromosomal translocation in T and B cell neoplasia, *Science* **234**:982–985.

Fukuhara, S., Hinuma, Y., Gotoh, Y.-I., and Uchino, H., 1983, Chromosome aberrations in T lymphocytes carrying adult T cell leukemia-associated antigens (ATLA) from healthy adults, *Blood* **61**:205–207.

Gale, R.P., and Rai, K., 1987, Chronic lymphocytic leukemia: Recent progress, future directions. Alan R. Liss: New York, N.Y.

Hayward, W.S., Neel, B.G., and Astrin, S.M., 1981, Activation of cellular onc gene by promoter insertion in ALV-induced lymphoid leukosis, *Nature* **290**:475–480.

Hecht, F., Kaiser McCaw, B., Peakman, D., and Robinson, A., 1975, Non-random occurrence of 7-14 translocations in human lymphocyte cultures, *Nature* **255**:243–244.

Hecht, F., Morgan, R., Hecht, B.K.-M., and Smith, S.D., 1984, Common region on chromosome 14 in T-cell leukemia and lymphoma, *Science* **226**:1445–1447.

Hecht, F., Morgan, R., Gemmill, R.M., Hecht, B.K.-M., and Smith, S.D., 1985, Translocations in T-cell leukemia and lymphoma, *N. Engl. J. Med.* **313**:758–759.

Hershfield, M.S., Kurtzberg, J., Harden, E., Moore, J.O., Whang-Peng, J., and Haynes, B.F., 1984, Conversion of a stem cell leukemia from a T-lymphoid to a myeloid phenotype induced by the adenosine deaminase inhibitor 2' deoxycoformicin, *Proc. Natl. Acad. Sci. USA* **81**:253–257.

Hollis, G.F., Mitchell, K.F., Battey, J., Potter, H., Taub, R.A., Lenoir, G., and Leder, P., 1984, A variant translocation places the lambda immunoglobulin genes 3' to the *c-myc* oncogene in Burkitt lymphoma, *Nature* **307**:752–755.

Isobe, M., Erikson, J., Emanuel, B.S., Nowell, P.C., and Croce, C.M., 1985, Location of gene for β subunit of human T-cell receptor at band 7q35, a region prone to rearrangements in T cells, *Science* **228**:580–582.

Jones, C., Morse, H.G., Kao, F.-T., Carbone, A., and Palmer, E., 1985, Human T-cell receptor α-chain genes: Location on chromosome 14, *Science* **228**:83–85.

Kaiser-McCaw, B., Hecht, F., Harnden, D.G., and Teplitz, R.L., 1975, Somatic rearrangement of chromosome 14 in human lymphocytes, *Proc. Natl. Acad. Sci. USA* **72:**2071–2075.

Kataoka, T., Kawakami, T., Takahashi, N., and Honjo, T., 1980, Rearrangement of immu-noglobulin γ1-chain and mechanism for heavy chain class switch, *Proc. Natl. Acad. Sci, USA* **77:**919–923.

Kennaugh, A.A., Butterworth, S.V., Hollis, R., Baer, R., Rabbitts, T.H., Taylor, A.M.R., 1986, The chromosome breakpoint at 14q32 in an ataxia-telangiectasia t(14;14) T cell clone is differ-ent from the 14q32 breakpoint in Burkitt's and an inv(14) T cell lymphoma. *Hum. Genet.* **73:**254–259.

Kirsch, I.R., Morton, C.C., Nakahara, K., and Leder, P., 1982, Human immunoglobulin heavy chain genes map to a region of translocation in malignant lymphocytes, *Science* **216:**301–303.

Kirsch, I.R., Brown, J.A., Lawrence, J., Korsmeyer, S., and Morton, C.C., 1985a, Translocations that highlight chromosomal regions of differentiated activity, *Cancer Genet. Cytogenet.* **18:**159–171.

Kirsch, I.R., 1985b, Burkitt's lymphomas translocate immunoglobulin and *c-myc* genes, in: *Genetic analysis of lymphoid neoplasms*, (T.A. Waldmann, ed.), *Ann. Int. Med.* **102:**497–510.

Klein, G., 1983, Specific chromosomal translocations and the genesis of B-cell-derived tumors in mice and men, *Cell* **32:**311–315.

Kohn, P.H., Whang-Peng, J., and Levis, W.R., 1982, Chromosomal instability in ataxia telangiec-tasia, *Cancer Genet. Cytogenet.* **6:**289–302.

Kurtzberg, J., Bigner, S.H., and Hershfield, M.S., 1985, Establishment of the DU.528 human lymphohemopoietic stem cell line, *J. Exp. Med.* **162:**1561–1578.

Land, H., Parada, L.F., and Weinberg, R.A., 1983, Cellular oncogenes and multistep carcinogen-esis, *Science* **222:**771–778.

LeBeau, M.M., Diaz, M.O., Rowley, J.D., and Mak, T.W., 1985, Chromosomal localization of the human T cell receptor β-chain genes, *Cell* **41:**335.

Leder, A., Pattengale, P.K., Kuo, A., Stewart, T.A., and Leder, P., 1986, Consequences of the widespread deregulation of the *c-myc* gene in transgenic mice: Multiple neoplasms and normal development, *Cell* **45:**485–495.

Leder, P., Battey, J., Lenoir, G., Moulding, C. Murphy, W., Potter, H., Stewart, T., and Taub, R., 1983, Translocations among antibody genes in human cancer, *Science* **222:**765–771.

Lewis, W.H., Michaelopoulos, E.E., Williams, D.L., Minden, M.D., and Mak, T.W., 1985, Breakpoints in the human T-cell antigen receptor α-chain locus in two T-cell leukemia patients with chromosomal translocations, *Nature* **317:**544–546.

Maki, R., Traunecker, A., Sakano, H., Roeder, W., and Tonegawa, S., 1980, Exon shuffling generates an immunoglobulin heavy chain gene, *Proc. Natl. Acad. Sci. USA* **77:**2138–2142.

Malcolm, S., Barton, P., Bentley, D.L., Ferguson-Smith, M.A., Murphy, C.S., and Rabbitts, T.H., 1982, Assignment of IgKV locus for immunoglobulin light chains to the short arm of chromosome 2 (2p13-cen) by *in situ* hybridization using a cRNA probe of HK101 γ Ch4A, *Cytogenet. Cell Genet.* **32:**296.

Mathieu-Mahul, D., Caubet, J.F., Bernheim, A., Mauchauffe, M., Palmer, E., Berger, R., and Larsen, C.-J., 1985, Molecular cloning of a DNA fragment from human chromosome 14 (14q11) involved in T cell malignancies, *EMBO J.* **4:**3427–3433.

McBride, O., Hieter, P., Hollis, G., Swan, D., Otey, M., and Leder, P., 1982, Chromosomal location of human kappa and lambda immunoglobulin light chain constant region genes, *J. Exp. Med.* **155:**1480–1491.

McFarlin, D.E., Strober, W., and Waldmann, T.A., 1972, Ataxia-telangiectasia, *Medicine* **51:**281–314.

Mengele, L., Willard, H.F., Smith, C.I.E., Hammerstrom, L., Fischer, P., Sherrington, P., Lucas, G., Thompson, P.W., Baer, R., and Rabbitts, T.H., 1987. Human T-cell tumors containing chromosome 14 inversion or translocation with breakpoints proximal to immunoglobulin join-ing regions at 14q32, *EMBO J.* **6:**2273–2280.

Miyamoto, K., Sato, J., Kitajima, K.-I., Togawa, A., Suemaru, S., Sanada, H., and Tanaka, T., 1983, Adult T-cell leukemia. Chromosome analysis of 15 cases, *Cancer* **52:**471–477.

Morton, C.C., Duby, A.D., Eddy, R.L., Shows, T.B., and Seidman, J.G., 1985, Genes for β chain of human T cell antigen receptor map to regions of chromosomal rearrangements in T cells, *Science* **228:**582–585.

Murre, C., Waldmann, R.A., Morton, C.C., Bongiavanni, K.F., Waldmann, T.A., Shows, T.B., and Seidman, J.G., 1985, Human γ-chain genes are rearranged in leukaemic T cells and map to the short arm of chromosome 7, *Nature* **316:**549–552.

Nepveu, A., Marcu, K.B., 1986, Intragenic pausing and anti-sense transcription within the murine *c-myc* locus, *EMBO J.* **5:**2859–2865.

Neri, A., Barriga, P., Magrath, I.T., Knowles, D.M., and Dalla-Favera, R., 1988, Different regions of the immunoglobulin heavy chain locus are involved in chromosomal translocations in endemic and sporadic forms of Burkitt lymphoma. *Proc. Natl. Acad. Sci. USA* (in press).

Nowell, P.C., and Hungerford, D.A., 1960, A minute chromosome in human chronic granulocytic leukemia, *Science* **132:**1197.

Ravetch, J.V., Kirsch, I.R., and Leder, P., 1980, Evolutionary approach to the question of immunoglobulin heavy chain switching: Evidence from cloned human and mouse genes, *Proc. Natl. Acad. Sci. USA* **77:**6734–6738.

Ravetch, J.V., Siebenlist, U., Korsmeyer, S., Waldmann, T., and Leder, P., 1981, Structure of the human immunoglobulin μ locus: Characertization of embryonic and rearranged J and D genes, *Cell* **27:**583–591.

Rowley, J.D., 1980, Ph¹-positive leukaemia, including chronic myelogenous leukemia, *Clin. Haematol.* **9:**55–86.

Sadamori, N., Kusano, M., Nishino, K., Tagawa, M., Yao, E.-I., Yamada, Y., Amagasaki, T., Kinoshita, K.-I., and Ichimaru, M., 1985, Abnormalities of chromosome 14 at band 14q11 in Japanese patients with adult T-cell leukemia, *Cancer Genet. Cytogenet.* **17:**279–282.

Scheres, J.M.J.C., Hustinx, T.W.J., and Weemaes, C.M.R., 1980, Chromosome 7 in ataxia-telangiectasia, *J. Pediatr.* **97:**440–441.

Seidman, J.G., and Leder, P., 1978, The arrangement and rearrangement of antibody genes, *Nature* **276:**790–795.

Shah-Reddy, I., Mayeda, K., Mirchandani, I., and Koppitch, F.C., 1982, Sezary syndrome with a 14:14 (q12:p31) translocation, *Cancer* **49:**75–79.

Shen-Ong, G.L.C., Keath, E.J., Piccoli, S.P., and Cole, M.D., 1982, Novel myc oncogene RNA from abortive immunoglobulin gene recombination in mouse plasma-cytomas, *Cell* **31:**443–452.

Showe, L.C., and Croce, C.M., 1987, The role of chromosomal translocations in B- and T-cell neoplasia, *Ann. Rev. Immunol.* **5:**253–277.

Smith, S.D., Morgan, R., Link, M.P., McFall, P., and Hecht, F., 1986, Cytogenetic and immunophenotypic analysis of cell lines established from patients with T-cell leukemia/lymphoma, *Blood* **67:**650–656.

Shimizu, A., and Honjo, T., 1984, Immunoglobulin class switching, *Cell* **36:**801–809.

Stewart, T.A., Pattengale, P.K., and Leder, P., 1984, Spontaneous mammary adenocarcinoma in transgenic mice that carry and express MTV/myc fusion genes, *Cell* **38:**627–637.

Taub, R., Kirsch, I., Morton, C., Lenoir, G., Swan, D., Tronick, S., Aaronson, S., and Leder, P., 1982, Translocation of the *c-myc* gene into the immunoglobulin heavy chain locus in human Burkitt lymphoma and murine plasmacytoma cells, *Proc. Natl. Acad. Sci. USA* **79:**7837–7841.

Taub, R., Moulding, C., Battey, J., Murphy, W., Vasicek, T., Lenoir, G., and Leder, P., 1984, Activation and somatic mutation of the translocated *c-myc* gene in Burkitt lymphoma cells, *Cell* **36:**339–348.

Taylor, A.M.R., 1982, Cytogenetics of ataxia-telangiectasia, in: *A cellular link between cancer, neuropathology, and immunodeficiency* (B.A. Bridges and D.G. Harnden, eds.) pp. 53–81,

John Wiley and Sons, New York.

Tsujimoto, Y., Finger, L.R., Yunis, J.J., Nowell, P., and Croce, C.M., 1984a, Cloning of the chromosome breakpoint of neoplastic B cells with the t(14;18) chromosome translocation, *Science* **226**:1097–1099.

Tsujimoto, Y., Yunis, J.J., Onorato-Showe, L., Erikson, J., Nowell, P.C., and Croce, C.M., 1984b, Molecular cloning of the chromosomal breakpoints of B cell leukemias and lymphomas with the t(11;14) chromosome translocation, *Science* **224**:2403–2406.

Tsujimoto, Y., Jaffe, E., Cossman, J., Gorham, J., Nowell, P.C., and Croce, C.M., 1985, Clustering of breakpoints on chromosome 11 in human B-cell neoplasms with the t(11;14) chromosome translocation, *Nature* **315**:340–343.

Ueshima, Y., Fukuhara, S., Hattori, T., Uchiyama, T., Takatsuki, K., and Uchino, H., 1981, Chromosome studies in adult T-cell leukemia in Japan: Significance of trisomy 7, *Blood* **58**:420–425.

Ueshima, Y., Rowley, J.D., Variakojis, D., Winter, J., and Gordon, L., 1984, Cytogenetic studies on patients with chronic T cell leukemia/lymphoma, *Blood* **63**:1028–1038.

Wake, N., Minowada, J., Park, B., Sandberg, A.A., 1982, Chromosomes and causation of human cancer and leukemia XLVII. T-cell acute leukemia in ataxia-telangiectasia, *Cancer Genet. Cytogenet.* **6**:345–357.

Welch, J.P., and Lee, C.L.Y., 1985, Non-random occurrence of 7-14 translocations in human lymphocyte cultures, *Nature* **255**:241–242.

Westbrook, C.A., Rubin, C.M., LeBeau, M.M., Kaminer, L.S., Smith, S.D., Rowley, J.D., Diag, M.O., 1987, Molecular analysis of TCR B and ABL in a t(7;9) containing cell lines (SUP-T3) from a human T cell leukemia, *Proc. Natl. Acad. Sci. USA* **84**:251–255.

Whang-Peng, J., Bunn, P.A., Knutsen, T., Kao-Shan, C.S., Broder, S., Jaffe, E.S., Gelmann, E., Blattner, W., Lofters, W., Young, R.C., and Gallo, R.C., 1985, Cytogenetic studies in human T-cell lymphoma virus (HTLV)-positive leukemia-lymphoma in the United States, *J. Natl. Cancer Inst.* **74**:357–369.

Williams, D.L., Look, A.T., Melvin, S.L., Roberson, P.K., Dahl, G., Flake, T., and Stass, S., 1984, New chromosomal translocations correlate with specific immunophenotypes of childhood acute lymphoblastic leukemia, *Cell* **36**:101–109.

Yoshikai, Y., Clark, S.P., Taylor, S., Sohn, U., Wilson, B., Minden, M.D., and Mak, T., 1985, Organization and sequences of the variable, joining, and constant region genes of the human T-cell receptor α-chain, *Nature* **316**:837–840.

Yunis, J.J., and Chandler (Harper), M.E., 1977, High-resolution chromosome analysis in clinical medicine, in: *Progress in Clinical Pathology*, Vol. 7 (M. Stefanini, A.A. Hossaini, and H.D. Isenberg, eds.), pp. 267–288, Grune and Stratton, New York.

Yunis, J.J., 1983, The chromosomal basis of human neoplasia, *Science* **221**:227–236.

Zech, L., Haglund, U., Nilsson, K., and Klein, G., 1976, Characteristic chromosomal abnormalities in biopsies and lymphoid-cell lines from patients with Burkitt and non-Burkitt lymphomas, *Int. J. Cancer* **17**:47–56.

Zech, L., Gahrton, G., Hammarstrom, L., Juliusson, G., Mellstedt, H., Robert, K.H., and Smith, C.I.E., 1984, Inversion of chromosome 14 marks human T-cell chronic lymphocytic leukemia, *Nature* **308**:858–860.

Ziegler, J.L., 1981, Burkitt's lymphoma, *N. Engl. J. Med.* **305**:735–745.

10

MHC-Disease Associations and T Cell-Mediated Immunopathology

ROLF M. ZINKERNAGEL

1. Introduction

Major transplantation antigens coded by the major histocompatibility gene complex (MHC) play a key role in lymphocyte interactions and in immunological recognition. It has become clear over the last 10 years that MHC-gene products (i.e., HLA antigens in humans, H-2 in mice) restrict T-cell specificity or guide T-cell function according to well-established rules (Zinkernagel and Doherty, 1979; Paul and Benacerraf, 1977; Townsend and McMichael, 1984; Möller, 1977); T cells recognise foreign antigens only on cell surfaces and only together with self-transplantation antigens. T-cell function is determined and T-cell responsiveness is regulated by the class of MHC antigens recognised as self. Class I MHC genes (HLA-A,B,C, or H-2K,D,L) regulate activities of class I-restricted, cytotoxic T cells; class II MHC genes (i.e., HLA-D or H-2I) regulate class II-restricted, differentiation-promoting T cells such as helper T cells or T cells involved in delayed type hypersensitivity.

Associations between major transplantation antigens and specific diseases are usually relatively weak but have been recognised for some time (Dausset and Svejgaard, 1977; McDevitt and Bodmer, 1974; Möller, 1983). It is still unclear

*Essential parts of this paper have been presented at the meeting of the European Association of Clinical Investigation in Toulouse, April, 1985 and are reproduced here with the permission of the European Journal of Clinical Investigation and Blackwell Scientific Publications, Ltd., Oxford, England.

ROLF M. ZINKERNAGEL ● Institute for Pathology, University Hospital Zurich, CH-8091 Zurich, Switzerland.

how these associations can be explained (Benecerraf and McDevitt, 1972; Ceppellini, 1973; Doherty and Zinkernagel, 1975; Jersild *et al.*, 1976; Blank and Lilly, 1977; Dupont *et al.*, 1977; Simons and Amiel, 1977; Helenius *et al.*, 1978; Kvist *et al.*, 1978; Geczy *et al.*, 1983; Kagnoff *et al.*, 1984; Parker *et al.*, 1979; White *et al.*, 1984). MHC-regulation of T cells controlling disease has been explained as follows: by (1) failure of infectious agents to associate with MHC antigens (altered self) recognisable by T cells (Doherty and Zinkernagel, 1975; Blank and Lilly, 1977; Helenius *et al.*, 1978; Kvist *et al.*, 1978); (2) holes in the receptor repertoire of T-cell recognition, which is influenced by MHC genes (Ir genes), (Benacerraf and McDevitt, 1972; Doherty and Zinkernagel, 1975); (3) molecular mimicry whereby infectious agents may resemble MHC antigens which cause tolerance (Geczy *et al.*, 1983; Kagnoff *et al.*, 1984); (4) MHC antigens possibly functioning more or less efficiently as receptor structures for infectious agents (Helenius *et al.*, 1978; Kvist *et al.*, 1978); (5) disturbed function of certain MHC-coded cell-surface antigens in differentiation processes, often causing lymphohematopoietic disorders (Dupont *et al.*, 1977; Simons and Amiel, 1977); (6) abnormalities of MHC coded complement components (Jersild *et al.*, 1976; Parker *et al.*, 1979) or (7) disfunction of genes not involved in immunology but closely linked to the MHC such as 21-hydroxylase (White *et al.*, 1984) or 6-GPD (Ceppellini, 1973). The present hypothesis is derived from experiments on T-cell-mediated immunopathology triggered by the noncytopathic lymphocytic choriomeningitis virus (LCMV) in mice (Hotchin, 1971; Lehmann-Grube, 1971; Pfau *et al.*, 1982; Ahmed *et al.*, 1984a; Allan and Doherty, 1985; Zinkernagel *et al.*, 1985), a virus that causes a disease spectrum resembling that of hepatitis B virus infections in man (Mackay, 1976). We propose here and provide supporting evidence that MHC-linkage of disease susceptibility may often signal T-cell-mediated and MHC-regulated immunopathology triggered by noncytopathic infectious agents.

2. Proposal

Many diseases, susceptibility to which is linked to the MHC, may reflect T-cell-mediated and MHC- (HLA or H-2) regulated immunological pathogenesis triggered by poorly or noncytopathic intracellular infectious agents. Acute cytopathic infectious agents do not usually show MHC-associated disease susceptibility since most members of the species must respond well immunologically and carry many other nonimmunological resistance genes. Poorly or noncytopathic intracellular infectious agents which often cause chronic disease and do not usually interfere with reproduction may establish various balanced states of host–parasite relationships. Besides many other factors, this balance crucially depends upon MHC-regulated T-cell immunity. It is influenced on the one side by the virulence and/or antigenic variation of the intracellular agent and on the other side by resistance, i.e., the general genetic make up of the host that is modulated by incidental environmental factors. In addition, these basic phenotypes may combine to allow MHC-coded genetically dominant immunoregulation to influence T-cell immunity and therefore the severity of disease, thus causing recognizable MHC-disease associations: strong T-cell immunity may result in relatively mild disease and in rapid elimination of the agent; weak T-cell immunity may lead to chronic infection with ensuing slow immunopathology caused by T-cell mediated tissue and cell damage or the triggering of autoantibody responses; and a complete absence of immunological defence may result in an often asymptomatic carrier state.

This proposal may help to explain why MHC-disease associations are relatively weak and variable. Explanations may include: first, infectious agents that by classical serology are apparently identical, may differ antigenically as recognised by T cells (Ahmed *et al.*, 1984a); second, many non-HLA genes that influence infections vary greatly throughout the outbred human population and third, HLA types defined by classical serology may actually comprise several variant HLA molecules that are recognised differently as self by T cells and therefore regulate responses differentially (Zinkernagel and Doherty, 1979; Biddison *et al.*, 1980; Cohen and Dausset, 1983). Our experiments with inbred mice and cloned virus isolates have reduced these numerous variables and show that under well-defined conditions MHC-disease associations may be very strong and directly reflect MHC-regulated T-cell-mediated immunopathology. This suggests: that infectious agents (known, suspected, or unknown) may be responsible for many of the MHC-linked diseases, not directly, but indirectly by triggering immunopathology; that such diseases may be preventable by appropriate vaccinations, and that successful therapeutic immune modulating intervention in such diseases depends upon exact knowledge of the host–parasite relationship.

3. Host–Parasite Relationship: General Parameters

The immune system of higher vertebrates is aimed at eliminating acute cytopathic intracellular and extracellular infectious agents and together with other host factors plays a major role in host resistance. The more acutely cytopathogenic an agent is, the more important is rapid and efficient protection by genetic resistance factors and immunologic defence. Poxvirus (Blanden, 1974; Fenner *et al.*, 1974), poliomyelitis virus (Salk, 1955), hepatitis A virus (Maynard, 1976), or the acute facultative intracellular bacterium *L. monocytogenes* (Hahn and Kaufmann, 1981) belong to this category. Because infections with these cytopathic agents are life threatening, particularly to the young, they may interfere with reproduction. Therefore, genetic resistance, including immunoprotection at the level of the species, is of paramount importance and every member of the species should be a high responder against such agents. This rule is well illustrated by the following exception, which was recently discovered. Laboratory mice are in general resistant to Sendai virus infections; one might expect that if MHC genes are allowed to mutate under highly protected laboratory conditions, MHC genes may emerge that cause low responsiveness to Sendai virus. Kast *et al*. have, in fact, found such examples amongst C57BL/6Kbm mutant mice (Kast, 1986). In contrast to these acutely cytopathic agents, noncytopathic viruses such as hepatitis B virus in humans or lymphocytic choriomeningitis virus (LCMV) in mice or the intracellular *Mycobacterium leprae* do not cause cell and tissue damage directly themselves. This is best illustrated by the fact that carriers lacking any immune response against these infectious agents are usually clinically well (hepatitis B, LCMV) and/or can survive for a long time (polar lepromatous leprosy). Cell and tissue damage by these types of infections are caused by immunological mechanisms.

Host resistance is regulated by a variety of genetic factors (Bang, 1973; Mims, 1982) including those controlling interferon levels (Goodman and Koprowski, 1962), macrophage activation (Friedman and Vogel, 1983), etc. In addition, many environmental factors such as nutrition, hygiene standards, superinfections, etc., may influence resistance; but quite obviously, MHC-dependent regulation of T-cell immunity must play a decisive role in some of these diseases, otherwise susceptibility to disease would not be linked to major transplantation antigens.

The parasite may influence disease by various mechanisms: virulence factors such as rapidity of replication (Pfau *et al*., 1982), susceptibility to interferons or other cellular resistance mechanisms (Mims, 1982; Friedman and Vogel, 1983), or preferential infection of particular cells or organs; all these characteristics may vary slightly from one virus isolate to another (Mims, 1982; Pfau *et al*., 1982; Ahmed *et al*., 1984). It has also become evident that otherwise serologically identical viruses may differ when analysed either with monoclonal antibodies, as shown for rabies virus (Wiktor and Koprowski, 1978), or by cloned T cells as demonstrated for LCMV (Ahmed *et al*., 1984a; Byrne *et al*., 1984). Therefore, viruses that differ antigenically in a way recognised by T cells but that is not registered by classical serology, may induce stronger or weaker T-cell responses, dependent upon the MHC of the host.

Because cytopathic agents such as poxvirus cause tissue damage directly, early and efficient elimination of the virus is mandatory for host survival. Virus-specific cytotoxic T cells apparently play a decisive role in stopping virus replication by destroying infected cells (Byrne *et al*., 1984; Zinkernagel and Althage, 1977). Alternatively, T cells destroy intracellular bacteria by way of activating macrophages (Mackaness, 1964). Thus, immune protection is essentially mediated by T-cell-mediated host-cell destruction or T-cell-dependent inflammatory reactions. Since T cells cannot distinguish cytopathic agents from poorly or noncytopathic ones, both trigger the same immunological mechanisms causing tissue damage with the aim of eliminating infectious agents (Zinkernagel, 1979).

Clinical disease initiated by noncytopathic agents is therefore caused by T-cell-mediated cytotoxicity and/or inflammatory mechanisms; severity depends mainly upon the balance between the spread of virus and the MHC-regulated immune response. The result in MHC-regulated high responders may be rapid elimination of the noncytopathic agent resulting in limited tissue damage, therefore causing more or less limited disease, rapid recovery, and good protection. In contrast, immunological unresponsiveness caused by MHC-linked nonresponder status with the same infection may lead to a carrier state which often does not cause disease, because there is no tissue destruction by the virus (e.g., hepatitis B virus, LCMV). In between are found the slowly reacting MHC-linked low responders, where T-cell-mediated cell and tissue damage becomes extensive because the noncytopathic agent has time to spread more widely; the result is more or less acute or chronic immunopathology (Zinkernagel, 1979) such as in aggressive hepatitis (Mackay, 1976).

4. Supporting Evidence

LCM disease in mice is a laboratory model used to study T-cell-mediated immunopathology (Hotchin, 1971; Lehmann-Grube, 1971; Zinkernagel et al., 1985; Allan and Doherty, 1985; Pfau et al., 1982). LCMV injected intracerebrally into immunocompetent mice leads to a severe lymphocytic choriomeningitis and death. In contrast, immunosuppressed mice do not die but become virus carriers (Hotchin, 1971; Lehmann-Grube, 1971; Zinkernagel and Doherty, 1979). It has been shown that T cells, particularly cytotoxic LCMV-specific T cells are responsible for LCM disease and death (Cole et al., 1972; Doherty and Zinkernagel, 1975; Zinkernagel and Doherty, 1979; Zinkernagel, 1979). LCMV injected ic. predominantly infects leptomeninges and it is most likely that effector T cells and recruited inflammatory cells destroy these infected leptomeningeal cells, thereby causing brain edema and death. This disease is virus-dose dependent and occurs only if there is a preferential recruitment of specific T cells to the brain rather than other infection sites in the body (Zinkernagel and Doherty, 1979; Hotchin, 1971; Pfau et al., 1982; Ahmed et al., 1984). LCM disease has been suspected to be associated with H-2 for some time (Oldstone et al., 1973). When two new LCMV isolates, A and D (Pfau et al., 1982), were tested, we found that inbred strains of mice differed vastly with respect to susceptibility to these two LCMV isolates; some mouse strains were susceptible to both LCMV-A and LCMV-D, others were resistant to both, and again others were susceptible to one LCMV isolate but resistant to the other (Zinkernagel et al., 1985). These differences can be taken to reflect the variation in genetic background genes found in outbred populations such as humans. All B10 mice tested were equally susceptible to LCMV-A; in contrast, the various B10 mouse strains sharing 99% of the genome and differing only in MHC genes showed a very strong influence of MHC genes on susceptibility to LCM disease caused by the LCMV-D isolate. Susceptibility to LCM disease was regulated by the class I MHC gene H-2D (corresponding to the human HLA-A locus); the D^q allele conferred susceptibility, the D^k, D^b, and D^d alleles conferred resistance. As in human MHC-disease associations, susceptibility was shown to be dominant in offsprings from susceptible and resistant parents. Analysis of LCMV-specific cytotoxic T-cell responses in B10 mice susceptible or resistant to LCM disease caused by LCMV-D established: the H-2D allele determines predominantly (i.e., to > 90%) the rapidity of onset and the activity of the cytotoxic T-cell response. Disease susceptibility correlated directly with early onset and high LCMV-specific and H-2D-restricted cytotoxic T-cell activity in both spleen and meningeal infiltrate cells. Theiler's murine encephalomyelitis virus (TMEV) which induces demyelinating disease in mice has recently also been shown to be influenced by H-2D-region genes in addition to non-MHC genes (Clatch et al., 1985); susceptibility correlates with TMEV-specific delayed type hypersensitivity.

The following human disease resembles the LCMV model closely: human hepatitis B virus infections may cause either an asymptomatic or clinically manifest disease resulting in recovery in about 95% of infected humans; or in an asymptomatic virus carrier state, in about 0.5%; or in chronic aggressive hepatitis, a clearcut immunopathological conflict situation, in about 2–5% of the patients. Susceptibility to the latter two phenotypes has been shown to be linked to HLA (Mackay, 1976; Jeannet and Farquet, 1974). Possibly, the spectrum of various states of disease caused by retroviruses such as the AIDS agent may also be an example of a delicate host–parasite relationship that could turn out to be regulated by the MHC. Lepra bacilli also cause asymptomatic or transient disease in greater than 90% of the population; in a few patients low and slow immune response will result in wide-spread, protracted, and therefore extensive inflammatory reactions causing nerve destruction (Bullock, 1978; Serjentson, 1983; de Vries et al., 1980). Furthermore, many HLA-linked diseases caused by autoantibodies (Vladutiu and Rose, 1971; Grabar, 1974; Rees et al., 1978; Carpenter, 1982) such as juvenile diabetes, autoimmune thyroiditis, rheumatoid disease, etc. may be explained along similar lines. Accordingly, T cells active against infectious agents may cause breakdown (Weigle, 1961) of B-cell tolerance by providing help to B cells specific for self-antigens (Weigle, 1961; Chiller et al., 1971) that are usually present in low concentrations (e.g., hormones, intracellular antigens). Viral infection of hormone-producing cells may induce virus-specific T-cells that may help B cells specific for self-antigens expressed on the same cell such as insulin, thyroglobulin, intrinsic factor, etc. B-cell tolerance to self-antigen depends on the concentration of the self-antigen and it has been convincingly documented that self-reacting B cells and autoantibodies are omnipresent, albeit in low concentrations (Grabar, 1974). In contrast, T-cell tolerance is practically absolute and truly self-reactive T cells have not been found (Weigle, 1961; Chiller *et al.*, 1971). Therefore, because helper T cells are required for most antibody responses, the amount of T help available and directed against foreign, e.g., viral or bacterial, antigenic determinants will determine whether self-specific B cells are triggered or not.

In all the examples discussed here the crucial T-cell response leading to disease is against nonself, against viral or bacterial antigens or against virally or bacterially altered self-transplantation antigens. Therefore from the point of view of the T-cell response, all these diseases are not autoimmune diseases but should be regarded as T-cell-mediated immunopathologies; some T cells cause tissue damage directly, others via triggering of bonafide autoimmune B cells.

Obviously this proposal demands an intensified search of causative infectious agents both known and unknown. This search has been demanded by many before, but becomes more imperative if one considers the immunological pathophysiology of these chronic, noncytopathic infections. Demonstration of the agent and its capacity to induce MHC-regulated immunopathology by T cells directly or by triggering of autoantibody producing B cells should be an ultimate proof.

Another attractive means of testing the proposal is through vaccination wherever the causative agent is known. The prediction is that it should generally prevent immunopathological disease. Vaccination against hepatitis B could prove an interesting model situation: does chronic aggressive hepatitis disappear if the host–parasite balance is influenced? Or does one merely shift the balance so that the now immunologically very poor responders that would have become hepatitis B carriers tend to develop chronic aggressive hepatitis, whereas those that originally would have had chronic aggressive hepatitis now will be free of immunopathological conflict situations? Similar considerations apply to lepra or the triggering of autoantibodies.

These examples, in addition, signal an important consequence of the proposal with respect to application of immune modulating therapeutic protocols during immunopathologically mediated disease. Unless one knows precisely the balance, kinetics, and actual relationship between parasite and host response, it is equally likely that one will influence the balance in a beneficial or a detrimental direction.

In conclusion, the proposal is made, illustrated, and supported by experimental and clinical evidence that T-cell-mediated immunopathology triggered initially by low- or noncytopathic infectious agents may cause chronic diseases that are linked to the MHC. It is obvious that not all MHC disease associations are explained by this pathophysiological mechanism, but a reasonable guess would be that many of such linkages may follow the outlined rules. Conversely, the proposal implies that MHC disease associations quite generally signal T-cell-mediated pathophysiology of the disease.

ACKNOWLEDGMENT. The research summarized here was supported by SNF 3.323-0.82, NIH AI-17285, and the Kanton of Zurich.

References

Ahmed, R., Byrne, J.A., and Oldstone, M.B.A., 1984a, Virus specificity of cytotoxic T lymphocytes generated during acute lymphocytic choriomeningitis virus infection: Role of the H-2 region in determining cross-reactivity for different lymphocytic choriomeningitis virus strains, *J. Virol.* **51**:34–41.

Ahmed, R., Salmi, A., Butler, L.D., Chiller, J.M., and Oldstone, M.B.A., 1984b, Selection of genetic variants of lymphocytic choriomeningitis virus in spleens of persistently infected mice, *J. Exp. Med.* **60**:521–540.

Allan, J.E., and Doherty, P.C., 1985, Consequences of a single Ir-gene defect for the pathogenesis of lymphocytic choriomeningitis, *Immunogenetics* **21**:581–590.

Bang, F.B., 1973, Genetics of resistance of animals to viruses, *Adv. Virus Res.* **23**:269–346.

Benacerraf, B., and McDevitt, H.O., 1972, Histocompatibility-linked immune response genes, *Science* **175**:273–279.

Biddison, W.E., Ward, F.E., Shearer, G.M., and Shaw, S., 1980, The self determinants recognized by human virus-immune T cells can be distinguished from the serologically defined HLA antigens, *J. Immunol.* **124**:548–555.

Blanden, R.V., 1974, T cell response to viral and bacterial antigens, *Transplant Rev.* **19**:56–84.

Blank, K.J., and Lilly, F., 1977, Evidence for an H-2 viral protein complex on the cell surface as the basis for the H-2 restriction of cytotoxicity, *Nature* **269**:808–810.

Byrne, J.A., Ahmed, R., and Oldstone, M.B.A., 1984, Biology of cloned cytotoxic T lymphocytes specific for lymphocytic choriomeningitis virus. I. Generation and recognition of virus strains and H-2b mutants, *J. Immunol.* **133**:433–440.

Bullock, W.E., 1978, Leprosy: A model of immunological perturbation in chronic infection, *J. Inf. Dis.* **137**:341–353.

Carpenter, C.B., 1982, Autoimmunity and HLA, *J. Clin. Immunol.* **2**:157–165.

Ceppellini, R., 1973, Old and new facts and speculations about transplantation antigens in man, *Prog. Immunol.* **1**:973–978.

Chiller, J.M., Habicht, G.S., and Weigle, W.O., 1971, Kinetic differences in unresponsiveness of thymus and bone marrow cells, *Science* **171**:813–815.

Clatch, R.J., Melvold, R.W., Miller, S.D., and Lipton, H.L., 1985, Theiler's murine encephalomyelitis virus demyelinating disease in mice is influenced by the H-2D region, *J. Immunol.* **135**:1408–1414.

Cohen, D., and Dausset, J., 1983, HLA-gene polymorphism, *Prog. Immunol.* **5**:1–12.

Cole, G.A., Nathanson, N., and Prendergast, R.A., 1972, Requirement of θ-bearing cells in lymphocytic choriomeningitis virus-infected central nervous system disease, *Nature* **238**:335–337.

Dausset, J., and Svejgaard, A., eds., 1977, *HLA and disease*, pp. 1–310, Munksgaard, Copenhagen.

Doherty, P.C., and Zinkernagel, R.M., 1975, A biological role for the major histocompatibility antigen, *Lancet* **i**:1406–1414.

Doherty, P.C., and Zinkernagel, R.M., 1975, T cell mediated immunopathology in viral infections, *Transplant Rev.* **19**:89–120.

Dupont, B., Good, R.A., Hauptmann, G., Schreuder, I., and Seligmann, M., 1977, Immunopathology, immunodeficiencies and complement deficiencies, in: *HLA and disease*, (Dausset J., Svejgaard A., eds.) pp. 233–248, Munksgaard, Copenhagen.

Fenner, F., McAuslan, B.R., Mims, C.A., Sambrook, J.F., and White, D.O., 1974, *The Biology of Animal Viruses*, Academic Press, New York.

Friedman, R.M., and Vogel, S.N., 1983, Interferons with special emphasis on the immune system, *Adv. Immunol.* **34**:97–140.

Geczy, A.F., Alexander, K., Bashir, H.V., Edmonds, J.P., Upfold, L., and Sullivan, J., 1983, HLA-B27, Klebsiella and ankylosing spondylitis, *Immunol. Rev.* **70**:23–50.

Goodman, G.T., and Koprowski, H., 1962, Macrophages as a cellular expression of inherited natural resistance, *Proc. Natl. Acad. Sci. USA* **48**:160–165.

Grabar, P., 1974, Self and not-self in immunology, *Lancet* **i**:1320–1323.

Hahn, H., and Kaufmann, S.H.E., 1981, The role of cell-mediated immunity in bacterial infections, *Ref. Infect. Dis.* **3**:1221–1250.

Helenius, A., Morein, B., Fries, E., and Simons, K., 1978, Human (HLA-A and HLA-B) and murine (H-2K and H-2D) histocompatibility antigens are cell surface receptors for Semliki Forest virus, *Proc. Natl. Acad. Sci. USA* **75**:3846–3850.

Hotchin, J., 1971, Persistant and slow virus infections, *Mongr. Virol.* **3**:1–140.

Jeannet, M., and Farquet, J.J., 1974, HLA antigens in asymptomatic chronic HBAg carriers, *Lancet* **ii**:1383–1384.

Jersild, C., Rubinstein, P., and Day, N.K., 1976, The HLA system and inherited deficiencies of the complement system, *Immunol. Rev.* **32**:43–71.

Kagnoff, M.F., Austin, R.K., Hubert, J.J., Bernardin, J.E., and Kasarda, D.D., 1984, Possible role for a human adenovirus in the pathogenesis of celiac disease, *J. Exp. Med.* **160**:1544–1557.

Kvist, S., Oesterberg, L., Persson, H., Philipson, L., and Peterson, P.A., 1978, Molecular association between transplantation antigens and cell surface antigen in adenovirus-transformed cell-line, *Proc. Natl. Acad. Sci. USA* **75**:5674–5678.

Lehmann-Grube, F., 1971, Lymphocytic choriomeningitis virus, *Virol. Monogr.* **10**:1–173.

Mackaness, G.B., 1964, The immunological basis of acquired cellular resistance, *J. Exp. Med.* **120**:105–120.

Mackay, I.R., 1976, The concept of autoimmune liver disease, *Bull. NY Acad. Med. Scil.* **52**:433–447.

Maynard, J.E., 1976, Hepatitis A., *Yale J. Biol. Med.* **49**:227–248.

McDevitt, H.O., and Bodmer, W.F., 1974, HLA immune response genes and disease, *Lancet* **i**:1269–1275.

Mims, C.A., 1982, *Pathogenesis of infectious disease* (2nd ed.), pp. 1–297, Academic Press, London.

Möller, G., (ed.), 1977, Ir-genes and T lymphocytes, *Immunol. Rev.* **38**:1–162.

Möller, G., (ed.), 1983, HLA and disease susceptibility, *Immunol. Rev.* **70**:1–180.

Oldstone, M.B.A., Dixon, F., Mitchell, G., and McDevitt, H.O., 1973, Histocompatibility linked genetic control of disease susceptibility: Murine lymphocytic choriomeningitis virus infection, *J. Exp. Med.* **137**:1201–1212.

Parker, K.L., Roos, M.H., and Shreffler, D.C., 1979, Structural characterization of the murine fourth component of complement and sex-limited protein and their precursors: Evidence for two loci in the S region of the H-2 complex, *Proc. Natl. Acad. Sci. USA* **76**:5853–5857.

Paul, W.E., and Benacerraf, B., 1977, Functional specificity of thymus-dependent lymphocytes, *Science* **195**:1293–1299.

Pfau, C.J., Valenti, J.K., Pevear, D.C., and Hunt, K.D., 1982, Lymphocytic choriomeningitis virus killer T cells are lethal only in weakly disseminated infections, *J. Exp. Med.* **156**:79–89.

Rees, A.J., Peters, D.K., Campston, D.A.S., and Batchelor, J.R., 1978, Strong association between HLA-DRW2 and antibody-mediated Goodpasture's syndrome, *Lancet* **i**:966–970.

Salk, J.E., 1955, A concept of the mechanism of immunity for preventing poliomyelitis, *Ann. NY Acad. Sci.* **61**:1023–1030.

Serjentson, S.W., 1983, HLA and susceptibility to leprosy, *Immunol. Rev.* **70**:89–112.

Simons, M.J., and Amiel, J.L., 1977, HLA and malignant diseases, in: *HLA and disease*, (Dausset, J., Svejgaard, A., eds.) pp. 212–232, Munksgaard, Copenhagen.

Townsend, A.R.M., and McMichael, A.J., 1984, The specificity of cytotoxic T lymphocytes stimulated with influenza virus; studies in mice and humans, *Progress in Allergy* **36:**10–97.

de Vries, R.R.D., Mehra, N.K., Vaidya, M.L., Gupte, M.D., Khan, P.M., and van Rood, J.J., 1980, HLA-linked control of susceptibility to tuberculoid leprosy and association with HLA-DR types, *Tissue Antigens* **16:**294–304.

Vladutiu, A., and Rose, N., 1971, Autoimmune murine thyroiditis. Relation to histocompatibility (H-2) type, *Science* **174:**1137–1139.

Weigle, W.O., 1961, The immune response of rabbits tolerant to bovine serum albumin to the injection of other heterologous serum albumins, *J. Exp. Med.* **114:**111–124.

White, P.C., New H.I., and Dupont, B., 1984, HLA-linked congenital adrenal hyperplasia results from a defective gene encoding a cytochrome P-45 specific steroid 21-hydroxylation, *Proc. Natl. Acad. Sci. USA* **81:**7505–7509.

Wiktor, T.J., and Koprowski, H., 1978, Monoclonal antibodies against rabies virus produced by somatic cell hybridization, detection of antigenic variants, *Proc. Natl. Acad. Sci. USA* **75:**3938–3943.

Zinkernagel, R.M., and Doherty, P.C., 1979, MHC-restricted cytotoxic T cells: studies on the biological role of polymorphic major transplantation antigens determining T cell restriction-specificity, function and responsiveness, *Adv. Immunol.* **27:**51–177.

Zinkernagel, R.M., Pfau, C.J., Hengartner, H., and Althage, A., 1985, A model for MHC disease associations: susceptibility to murine lymphocytic choriomeningitis maps to class I MHC genes and correlates with LCMV-specific cytotoxic T cell activity, *Nature* **316:**814–817.

Zinkernagel, R.M., and Althage, A., 1977, Antiviral protection by virusimmune cytotoxic T cells: Infected target cells are lysed before infectious virus progeny is assembled, *J. Exp. Med.* **145:**644–651.

Zinkernagel, R.M., 1979, Associations between major histocompatibility antigens and susceptibility to disease, *Ann. Rev. Microbiol.* **33:**201–213.

11

The γ-δ Heterodimer
A Second T-Cell Receptor?

NICOLETTE CACCIA, YOSHIHIRO TAKIHARA, and TAK W. MAK

T cells play an important role in the mammalian immune system, participating in a number of cell–cell interactions. It is through these interactions, and the products that result from these interactions, that T cells aid in the regulation and differentiation of the immune response. T cells can be divided into those that mediate the cellular response (cytotoxic T lymphocytes) and those that regulate the humoral response (helper and suppressor T lymphocytes). Cytotoxic T lymphocytes (CTL) lyse abnormal host cells, including neoplastic cells and those infected by virus, while helper T lymphocytes (HTL) enhance the response of B cells and other T cells, and suppressor T cells down-regulate their response. T cells are closely related to B cells, which provide the fine tuning of the immune response by the production of immunoglobulins; antigen-specific molecules involved in a variety of immune reactions that lead to the elimination of antigen and the neutralization of antigen-bearing cells (Davies and Metzger, 1983; Honjo, 1983). Both B and T cells have diverse, clonally distributed repertoires, recognize antigen in a specific manner by means of cell surface receptors, and generate the genes that encode their receptors by the rearrangement of sequences that are noncontiguous in germ-line DNA, using the same enzyme system. However, there are a number of significant differences between the two cell types, including their products, specific functions, and modes of antigen recognition. For example, although both B and T cells recognize antigen in a specific manner, T cells are unique in that they seem only to recognize antigen in the context of self major histocompatibility complex (MHC) antigens, a phenomenon

NICOLETTE CACCIA, YOSHIHIRO TAKIHARA, and TAK W. MAK ● The Ontario Cancer Institute, and Department of Medical Biophysics, University of Toronto, Toronto, Ontario, M4X 1K9 Canada.

known as MHC restriction (Zinkernagel and Dougherty, 1974). The high degree of similarity between the two cell types can be explained by the fact that over 98% of the genes expressed in the two cell types are the same. Thus, the molecules responsible for the functions specific to each cell type will be encoded by the remaining 2% of messages, which are expressed only in that cell type.

The group of T-cell-specific molecules has been well studied over the past few years, and includes such molecules as T4, T8, and the chains of the T-cell receptor. The T-cell receptor (TcR) is composed of an α chain disulphide linked to β chain, which together are responsible for antigen recognition (Haskins *et al.*, 1984; Kappler *et al.*, 1983; MacIntyre and Allison, 1983; Meuer *et al.*, 1983; Ohashi *et al.*, 1985). This dimer is associated with the chains of the CD3 complex, which are responsible for the transduction of the antigen-binding signal through the cell membrane to produce the appropriate T-cell function (Borst *et al.*, 1983; Weiss and Stobo, 1984; Weiss *et al.*, 1986b). The T-cell receptor has been studied intensely since the recent isolation of the genes that encode the α and β chains (Chien *et al.*, 1984a,b; Hedrick *et al.*, 1984a,b; Saito *et al.*, 1984a; Sim *et al.*, 1984; Yanagi *et al.*, 1984, 1985) and a number of interesting questions of immunology are being addressed using these genes.

1. Cloning of the γ Chain

In the course of the isolation of the α chain of the T-cell receptor, another interesting T-cell-specific molecule was isolated and named the γ chain (Saito *et al.*, 1984b). Like the α and β chains of the T-cell receptor, this gene is com posed of separate, noncontiguous coding segments in the genome, which undergo rearrangement in T cells to produce a complete gene. The γ chain is a member of the immunoglobulin supergene family, with a two domain structure that contains an internal disulphide bond in each domain. The amino terminal domain has the structure of a variable region, while the membrane-proximal domain has that of a constant region. On the basis of the early observation that the γ chain was primarily expressed in cytotoxic T cells, which, for the most part, recognize antigen in the context of class I MHC products, it was postulated that the γ chain was involved in the recognition of MHC by these cells (Heilig *et al.*, 1985; Kranz *et al.*, 1985b). This hypothesis has been weakened by several findings. The γ chain has been shown to be expressed in helper and autoreactive T cell clones (Zauderer *et al.*, 1986). The level of γ-chain expression in mature T cells is very low (Yoshikai *et al.*, 1987) and the majority of these messages seem to be nonfunctional (Reilley *et al.*, 1986; Rupp *et al.*, 1986; Yoshikai *et al.*, 1987). In addition, certain cytotoxic cells bearing the γ chain on their surface exhibit non-MHC restricted activity (Bluestone *et al.*, 1987; Borst *et al.*, 1987; Budd *et al.*, 1986; Pardoll *et al.*, 1987), suggesting that antigen recognition by γ-chain-containing complexes may be in a very different context than that of the TcR αβ chain complex.

2. Genomic Organization of the γ-Chain Loci in Mouse and Man

The human γ-chain locus is found on the short arm of chromosome 7 (Murre et al., 1985) and the murine on chromosome 13 (Kranz et al., 1985a). The basic genomic organization of the γ chain is reminiscent of those of the T-cell receptor α and β chains, and of the immunoglobulin genes, with a number of variable (V), joining (J), and constant (C) region gene segments (Hayday et al., 1985). The number of variable regions, while not as limited as has been thought originally, is much smaller than that of those associated with the T-cell receptor α and β loci (Garman et al., 1986; Hayday et al., 1985; Heilig and Tonegawa, 1986; Iwamoto et al., 1986;). However, there are more γ-chain constant-region genes than are found in the classical TcR loci, which have one α-chain and two β-chain constant-region genes. Schematic diagrams of the human and murine γ-chain loci are shown in Figs. 1 and 2, respectively.

2.1. The Murine γ-Chain Genes

Analysis of the genomic organization of the γ-chain genes in BALB/c mice revealed three cross-hybridizing constant regions ($C_\gamma 1$, $C_\gamma 2$, $C_\gamma 3$), each associated with its own J_γ segment (Hayday et al., 1985; Iwamoto et al., 1986) (see Fig. 2). Although these three C_γ segments are highly homologous, there are important differences between them. In Balb/c mice, $C_\gamma 1$ has one potential N-glycosylation site, while $C_\gamma 2$ has none, but both can code for functional proteins, and $C_\gamma 3$ is a pseudogene (Born et al., 1986; Hayday et al., 1985; Iwamoto et al., 1986; Saito et al., 1984b). Each of these C_γ genes consists of three exons, with a cysteine residue encoded by the second exon. Analysis of cDNA clones from cytotoxic T-cell clones, which were derived from the mouse strain C57BL/10 (B10), revealed the use of additional J_γ and C_γ ($C_\gamma 4$) segments in one of these clones (Iwamoto et al., 1986). These new sequences are located 5' to the inverted $V_\gamma(V_\gamma 1)$, in the germline of both B10 and BALB/c mice, and are in the same orientation as the inverted V_γ ($V_\gamma 1$). The $C_\gamma 4$ sequence shows only 60–70% homology to the previously isolated C_γ segments, suggesting that it may be an isotype with a different function. The genomic sequence of $C_\gamma 4$ is capable of encoding a functional message, has one potential N-glycosylation site and a lysine residue is found in the transmembrane region of the deduced protein sequence (Iwamoto et al., 1986). The presence of a positively charged lysine residue in the normally hydrophobic transmembrane region in the α and β chains of the T-cell receptor has been postulated to be important in their association

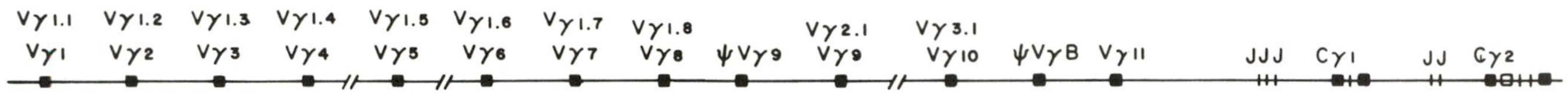

Figure 1. A schematic diagram of the genomic human γ chain locus. As can be seen there are a number of variable (V_γ) gene segments, which can be divided into four families, on the basis of sequence homology. Downstream of these V_γ, are two constant regions ($C_\gamma 1$ and $C_\gamma 2$), each of which is associated with its own set of joining gene segments (J_γ).

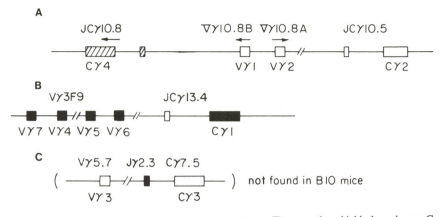

Figure 2. A diagram of the genomic murine γ chain locus. There are three highly homologous C_γ chain genes ($C_\gamma 1$, $C_\gamma 2$, $C_\gamma 3$), each associated with a cluster of variable (V_γ) and joining (J_γ) gene segments. A fourth C_γ gene ($C_\gamma 4$), which is not very homologous to the other three, is also found with associated J_γ and V_γ sequences. The genes shown in A and B are found in all mice strains studied to date, whereas the genes shown in C ($V_\gamma 3$, $J_\gamma 2.3$, and $C_\gamma 3$ (pseudogene)) are not found in certain strains of mice, notably C57BL/B10.

with the CD3 proteins, by the neutralization of the aspartic acid residue found in the transmembrane region of the CD3 components that span the membrane. It is possible that the lysine encoded by this γ-chain gene plays a similar role in a heterodimer. All the murine C_γ genes encode an extramembranous cysteine, indicating that they are capable of forming disulphide bonds with another protein to form a dimer.

The finding of new members of the murine γ-gene family, with low homology to previously isolated members, raises the possibility of the isolation of further members, and reveals that the repertoire of these genes is not as limited as was originally thought. Further characterization of the C56BL/10 genomic structure revealed that $V_\gamma 3$, $J_\gamma 3$, and the pseudogene $C_\gamma 3$, which are found in the BALB/c genome, are absent in that of C57BL/10 (Iwamoto *et al.*, 1986). There are several additional V_γ ($V_\gamma 4$, $V_\gamma 5$, $V_\gamma 6$, $V_\gamma 7$) segments 5' of $C_\gamma 1$. A complete picture of the murine locus is shown in Fig. 2 and an alignment of the deduced amino acid sequences of the murine V_γ genes is shown in Fig. 3.

2.2. The Human γ-Chain Genes

The human γ-chain repertoire would seem to be larger than that of the mouse, since a larger number of variable region genes have been identified by

```
Matches

6/6    G                                         HMY K          C                        L                            D   YYC
5/6    G                                         IHMY K        ISC                      E L                         TS L I        EDEATYYCA W
4/6    G G LEQ ELSVTR  DETAQISC IV   F NT      IHWYRQK  Q     E L YV            L  K KK EA SKDF    STS L IN L      INYLKKEDEATYYCAVW
3/6
2/6    GLGQLEQTELSVIREIDETAQISC IVSLPYFSNTAIHWYRQKPGQ L  EYLIYVSTNYNQRPLDGK KKIEASKDFQDSTSTLEINYLKKEDEATYYCAVWI


1    GLGQLEQTELSVTRETDFSAQISC IVSLPYFSNTAIHWYRQKAKKF  EYLIYVSTNYNQRPLGGKNKKIEASKDFQTSTSTLKINYLKKEDEATYYCAVWI
2    GLGQLEQTELSVTRETDENVQISC IVYLPYFSNTAIHWYRQKTNQQF EYLIYVATNYNQRPLGGKHKKIEASKDFKSSTSTLEINYLKKEDEATYYCAVWM
3                                                                                     DFQTSTSTLEINYLKKEDEATYYCAVWI
4    GHGKLEQPEISISRPRDETAQISC KVFIESFRSVTIHWYRQKPNQGL EFLLYVLATPTHIFLDKEYKKMEASKNPSASTSILTIYSLEEEDEAIYYCSY
5    GDSWISQDQLSFTRRPNKTVHISC KLSGVPLHNTIVHWYQLKEG     EPLRRIFYGSVKTYKQDKSHSRLEIDEKDDGTFYLIINNVTSDEATYYCACW
6    GTSLTSPLGSYVIKRKGNTAFLKCQIKTSVQKPDAYIHWYQEKPCQRLQRMLCSSSKETIVYEKDFSDERYEARTWQSDLSSVLTIHQVTEEDTGTYYCACW
```

Figure 3. An alignment of the deduced amino acid sequences of the murine Vγ genes. 1, Vγ1; 2, Vγ2; 3, Vγ3; 4, Vγ4; 5, Vγ5; 6, Vγ6. Spaces have been introduced to maximize regions of homology. Sequences 1–3 were derived from the germline V$_{2q}$ sequences from Hayday *et al.*, 1985, after removal of the introns. Sequences 4–6 are from the thymocyte cDNAs of Garman *et al.*, 1986. In all cases the leader sequences have been removed.

mapping (see Fig. 1) and Southern blot analysis (see Fig. 4) (LeFranc *et al.*, 1986c; Yoshikai *et al.*, 1987). These V_γ can be divided into three or four families (Yoshikai *et al.*, 1987). Two constant region genes have been isolated, which are highly homologous to each other in both the coding and 3′ untranslated regions (LeFranc and Rabbitts, 1985; LeFranc *et al.*, 1986a,b; Murre *et al.*, 1985) and it is possible that there are other human constant region genes that do not cross-hybridize to these, as in the case of $C_\gamma 4$ in the mouse system. Both C_γ genes have a structure similar to that of the C_α and C_β genes, with an extracellular constant domain followed by transmembrane and cytoplasmic portions and a lysine residue in the transmembrane region in an analogous position to those in the α and β chain genes and in the murine $C_\gamma 4$ gene. The human $C_\gamma 2$ gene contains 5 potential N-glycosylation sites (Asn-X-Ser-Thr) (Dialynas *et al.*, 1986), while the $C_\gamma 1$ gene encodes four (Krangel *et al.*, 1987). The human $C_\gamma 1$ gene has three unique exons (LeFranc *et al.*, 1986c). In contrast, duplication, and even triplication of the second exon of $C_\gamma 2$ (C_γII) have been described (Krangel *et al.*, 1987; LeFranc *et al.*, 1986b; Littman *et al.*, 1987). There are distinct differences in the amino acid sequence encoded by the different copies of C_γII, and all but one nucleotide change results in an amino acid substitution, suggesting that the divergence may be selectively encouraged (Krangel *et al.*, 1987) Comparison of $C_\gamma 2$ containing cDNA clones from different cell lines (Dialynas *et al.*, 1986; Krangel *et al.*, 1987; LeFranc *et al.*, 1986b) suggest that different transcripts may use different combinations of the different available C_γII exons to generate different γ chain proteins. Both exons I and III are highly conserved between $C_\gamma 1$ and $C_\gamma 2$, but exon II shows considerable divergence. The most striking result of this divergence is that $C_\gamma 1$ encodes a cysteine residue and $C_\gamma 2$ does not, which means that only $C_\gamma 1$ can participate in disulphide bonds. This is borne out by investigation of complexes involving $C_\gamma 2$, which show that they consist of γ monomers associated with CD3.

Alignment of the deduced amino acid sequences of the human V_γ genes is illustrated in Fig. 5. Each V_γ segment is labelled with the appropriate family assignment. The partial sequences of $V_\gamma 3$ and $V_\gamma 4$ genes are also included for comparison. Analysis of 17 human T-cell leukemia lines with V_γ probes suggests a germline organization of 5′ $V_\gamma 1$-$V_\gamma 2$-$V_\gamma 3$-J-$C_\gamma 3$ (Kimura *et al.*, 1988; LeFranc *et al.*, 1987).

3. γ-Chain Expression

One of the most unusual aspects of the γ chain is the high frequency of nonfunctional messages. In the human system, examination of six γ-chain cDNA sequences from thymocytes and 14 cDNA sequences from peripheral blood T cells revealed that all the transcripts were nonfunctional, either as the result of frameshift mutations, insertions, or deletions (Yoshikai *et al.*, 1987).

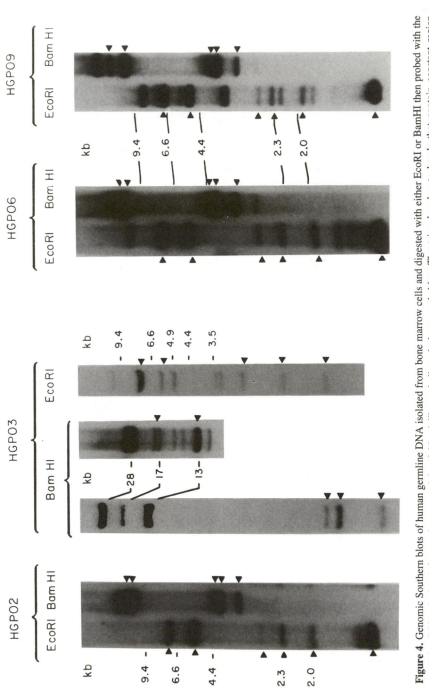

Figure 4. Genomic Southern blots of human germline DNA isolated from bone marrow cells and digested with either EcoRI or BamHI then probed with the γ- chain cDNAs isolated from a human peripheral blood library indicated above each blot. The triangles denote bands that contain constant region sequences. Molecular sizes are provided for reference beside each blot.

Figure 5. An alignment of the deduced amino acid sequences of the human Vγ genes.

Although functional γ-chain transcripts have been found in functionally active murine cytotoxic T lymphocytes (Jones *et al.*, 1986), the frequency of nonfunctional γ-chain transcripts is also very high (Reilly *et al.*, 1986; Rupp *et al.*, 1986), suggesting that, in certain compartments, there may be an active or passive process that selects against T cells that produce functional γ-chain transcripts. Recent studies have shown that, while there is a high proportion of nonfunctional γ-chain messages in thymocytes and mature T cells, functional γ-chain mRNA is produced by certain T-cell populations. For example, in young (8-week) athymic mice, which produce undetectable levels of full-length α- and β-chain messages (Kishihara *et al.*, 1987), the level of γ-chain expression is elevated compared to that found in normal mice, and the majority of the transcripts are full length (Yoshikai *et al.*, 1986). Analysis of cDNA library derived from these mice showed that 4/4 (2 $C_\gamma 1$ and 2 $C_\gamma 2$) γ transcripts isolated were functional, implying that the majority of full-length γ-chain transcripts in these mice are functional (Yoshikai *et al.*, 1986). This finding raises the possibility that Thy1[+] spleen cells found in athymic mice use these genes as recognition receptors.

If one looks at the expression of T-cell receptor messages in the murine thymus during embryogenesis, γ-chain mRNA appears first, followed by β-chain, and then α-chain messages (Haars *et al.*, 1986; Raulet *et al.*, 1985; Snodgrass *et al.*, 1985a,b). γ-chain expression levels are highest in early thymocytes (Haars *et al.*, 1986), declining slowly from day 16 until birth (Born *et al.*, 1985, 1986) and lowest in more mature thymocytes (Haars *et al.*, 1986). The frequency of nonfunctional messages in thymocytes is also high (Yoshikai *et al.*, 1987). Fetal thymocytes from stages early in development have been shown to use preferentially the $C_\gamma 1$ segment (Lew *et al.*, 1986).

In normal murine fetal thymus, the levels of γ-chain transcription is high (Raulet *et al.*, 1985), but it would seem that the majority of these messages are nonfunctional (Heilig and Tonegawa, 1986). However, in adult murine thymus, a small percentage of transcripts derived from double negative (CD4[-]CD8[-]) cells were functional. These are derived predominantly from the $C_\gamma 1$ locus (Lew *et al.*, 1986; Nakashini *et al.*, 1987). The double negative cells of fetal thymus also express high levels of $C_\gamma 1$ derived γ transcripts (Lanier and Weiss, 1986; Pardoll *et al.*, 1987; Weiss *et al.*, 1986a).

During murine fetal ontogeny, the CD3 molecule is associated with a disulphide linked γδ heterodimer as early as day 15, but by day 17.5, when most thymocytes are double positive and the CD4+CD8[-] population has begun to appear, there are equivalent amounts of γδ and αβ associated with CD3 with the γδ heterodimer on double negative cells, and the αβ on CD4+CD8[-] cells. By day 20, anti-CD3 only coprecipitates with αβ and an anti-γ antiserum only precipitates small amounts of the 35-kd/45-kd q/w heterodimer (Pardoll *et al.*, 1987). The γ chains in these fetal thymocytes are N glycosylated, and, as $C_\gamma 2$

has no potential N-glycosylation sites, $C_\gamma 3$ is a pseudogene and $C_\gamma 4$ is not detectable in fetal thymocytes (Heilig and Tonegawa, 1986); these γ transcripts are likely derived from either $C_\gamma 1$ or an as yet unidentified C_γ segment. In the day-16 fetal thymocytes that are CD3$^-$, there is no α-chain expression, but moderately high levels of β-chain expression and high levels of γ-chain expression, while in CD3$^+$ cells, there is no detectable α-chain expression and little or no β-chain expression, but high levels of γ expression, with concomitant expression of γ on the cell surface (Bluestone et al., 1987). The γ on the cell surface of these cells is 35 kd and is disulphide linked to a 45K protein (δ) (Lew et al., 1986; Nakashini et al., 1987). Cell sorting of adult thymocytes shows that most double negative cells express CD3, and of these, those that stain dully with anti-CD5 express cell-surface γ, and those that are CD5-bright have high levels of CD4 (Bluestone et al., 1987). However, the majority of adult thymocytes are double positive cells that express only low levels of CD3, so the γ-producing cells comprise only a small percentage of thymocytes and thus their presence may be overlooked in standard assays. In peripheral blood cells, there are only very low levels of γ expression. These transcripts are mainly $C_\gamma 2$ derived and, on the basis of cDNA analyses are most likely nonfunctional (Reilly et al., 1986; Rupp et al., 1986; Yoshikai et al., 1986). However, if peripheral blood T cells are sorted on the basis of expression of CD3 and β chain, and the CD3$^+$β$^-$ cells examined, a different picture emerges. IL-2 dependent CD3$^+$4$^-$8$^-$ T cell lines derived from these cells seem to be functionally competent, exhibiting a non-MHC restricted cytotoxicity, and do not express either α or β chain, but do express γ (Lanier et al., 1987). These cells compose up to 3% of peripheral blood T cells (0.5% of thymocytes) (Lanier et al., 1987), and are for the most part double negative.

The expression of γ can also be induced in certain populations. Jones et al. (1986) have shown that γ can be induced in mixed lymphocyte reactions (MLR). These cells have comparable levels of α, β, and γ chain RNA expression and the γ transcripts, derived from the $C_\gamma 2$ gene, are likely functional, as there is surface γ expression. When Maeda et al. (1987) examined MLR generated by allogeneic stimulation, they found two types of γ-chain protein on the surface of responder cells. The majority of cells seem to have $C_\gamma 1$-derived products, while a minority appear to use the $C_\gamma 2$ gene to produce γ chains. Miescher et al. (1987) have shown γ induction, by PMA and IL-2, in MRL 1pr/1pr mice double negative lymph-node cells. The transcripts are $C_\gamma 1$ derived and as the same percentage (18%) of these cells are KJ16$^+$ as those in the periphery of normal mice. Double positive (CD4$^+$CD8$^+$γ$^-$) thymic cells can also be induced under the same conditions (PMA and IL-2) to become γ$^+$ double negative cells (R. Spolski, personal communication). This raises the possibility that double positive cells, which form the majority of thymocytes, could act as a reserve of cells to be drawn upon and induced as the need for them arose, however, the testing of this hypothesis requires a great deal more information about T-cell ontogeny and the roles that α, β, and γ play in this process, than we have at our disposal at present.

It is possible that the γ chain serves an important function early in T-cell ontogeny or in a subset of T cells that we have not characterized. A diagram of a possible scheme of T-cell ontogeny is shown in Fig. 6. In the human system, small populations of double negative cells can be found in both peripheral blood (Borst *et al.*, 1987; Brenner *et al.*, 1987; Lanier *et al.*, 1987) and thymus (De-LaHera *et al.*, 1985; Toribo *et al.*, 1986) and it is in these populations that γ would seem to play an active role. The γδ receptor is also found on dendritic epidermal cells (Boryhadi *et al.*, 1987), suggesting that this receptor may play a role in the first line of defense mounted by these cells. The findings that γ expression is exceptionally high in a small population of thymocytes, the dou-

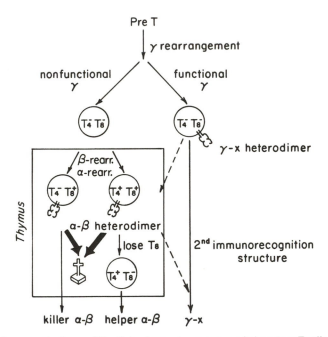

Figure 6. A proposed scheme of T-cell development. γ rearranges in immature T cells well before α or β. If this rearrangement proves non-functional, a cell can then go on to attempt rearrangement of the α and β chain genes. If these rearrangements prove successful the cell can then undergo thymic selection and either be eliminated or emerge from the thymus as a functional killer or helper cell. On the other hand, if the γ chain rearrangement produces a functional gene, the cell can go onto become a γ^+X^+ cell, following a path of differentiation, that may include a sojourn in the thymus and selection procedure. X is identified as δ or $C_\gamma 4$.

Table I. Cells Expressing γ-δ Receptor

Day 15–18 fetal thymocytes	Thy-1 + dendritic epidermal cells
CD4⁻ CD8⁻ thymocytes	Athymic mice Thy-1⁻ cells
1–5% peripheral T cells	MRL-1pr CD4⁻ CD8⁻ lymph node
Natural killer cells	cells + PMA, IL2

ble-negative cells (T4⁻T8⁻), and that productive γ-chain expression can be induced in certain populations under specific conditions, suggest that the expression of a functional cell-surface γ-chain molecule may be a state-dependent event that occurs only within a given microenvironment. Since there would seem to be different requirements for proliferation for thymocytes than for mature T cells (Bluestone *et al.*, 1987), this idea merits further study. The high level of coding sequence conservation between the γ genes suggests that the γ chain may play an important role in cells, but the nature of that role is only beginning to be investigated. A list of the cells expressing the γδ heterodimer on their cell surface is given in Table I.

4. γδ as a Possible Second T-Cell Receptor

One of the most interesting areas of research, in this area, is the isolation of T cells expressing surface γ chain in conjunction with the chains of the CD3 complex and another T-cell receptor molecule, the δ chain. On the large majority of T cells, the CD3 chains are closely associated with the α and β chains of the T-cell receptor, and serve to transduce the signal derived from the binding of the receptor to its target (Weiss *et al.*, 1986b). Binding of an antibody directed against the CD3 complex also brings down the T-cell receptor dimer, so one would expect that an antibody against the framework of the T-cell receptor would bind to the same population as would an anti-CD3 monoclonal antibody. This, however, is not the case. An anti-CD3 monoclonal has been shown to bind to a small population of cells in normal peripheral blood, that bind CD3, but not the TcR framework antibodies. Analysis of this population, obtained in stimulation of T lymphocytes from an immunodeficient patient, revealed that the CD3 molecule was associated with a novel heterodimer composed of a 55-Kd chain (which reacts with anti-γ antisera), which is noncovalently associated with a 40-kd chain, termed δ (Brenner *et al.*, 1986). Bank *et al.* (1986) isolated double-negative (CD4⁻CD8⁻) cells and, upon culture of these cells with IL-2 and feeder cells, saw high levels of γ expression with a concomitant expression of a nondisulphide linked 62-kd/44-kd dimer associated with CD3 on the cell surface, while Weiss *et al.* (1986a) saw surface γ expression on association with

CD3 on a CD3+α−β− leukemic cell line, but only as a single 56- to 60-kd chain.

Further study has shown that surface γ chain on human cells, is found in at least two forms. In the first, as described in Brenner *et al.* (1986, 1987) a 55-kd glycosylated (40-kd unglycosylated) γ is noncovalently associated with a non-γ protein in conjunction with CD3. The second form is described by Borst *et al.* (1987) and consists of a 70- to 80-kd disulphide linked dimer associated with CD3. This dimer can be divided into a 43-kd non-γ protein and 40-kd and 36-kd proteins that would seem to be differentially glycosylated forms of γ. γ would seem to exist in two very different forms on the human cell surface. The larger (55-kd) glycosylated form has only been found in a nondisulphide linked form (Brenner *et al.*, 1987), while the smaller 36- to 40-kd glycosylated peptide has been found in both disulphide linked and noncovalently associated forms.

Thus, the human γ chain can be differentially glycosylated to form either a 36-kd or a 40-kd protein. These two molecules can then either participate in the formation of a γγ homodimer (Brenner *et al.*, 1987; Moingeon *et al.*, 1987) or associate with a 43-kd non-γ protein (Borst *et al.*, 1987; Brenner *et al.*, 1987; Lanier *et al.*, 1987). The γ chain may also form a noncovalently linked hetero-dimer (Brenner *et al.*, 1986; Brenner *et al.*, 1987) or exist as a 40- to 44-kd monomer (Borst *et al.*, 1987; Brenner *et al.*, 1987; Lanier *et al.*, 1987).

The murine γ chain is expressed on the cell surface of CD3+γ+ cells as a 35-kd molecule disulphide linked to a 45-kd non-γ chain (δ)(Lew *et al.*, 1987b). This heterodimer is noncovalently linked to the CD3 complex.

It would seem that the δ chain genes are imbedded in the α chain locus, between the V_α segments and the J_α genes (see Figure 7). Comparison of puta-tive amino acid sequences derived from genes in this region to the peptide se-

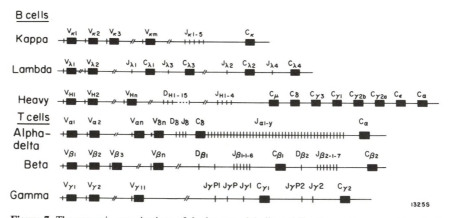

Figure 7. The genomic organizations of the immunoglobulin and T-cell receptor genes. These loci are composed of separate noncontiguous gene segments that undergo rearrangement during ontogeny to produce a mature antigen receptor. The δ-chain locus is unique in that it is nested in the α-chain locus.

quences of the δ chain protein, indicate that this region encodes the δ chain (Born *et al.*, 1987), as does the finding that monoclonal antibodies generated against synthethic peptides encoded by this region can precipitate γδ dimer from cells (Loh *et al.*, 1987). In mice, there is one C_δ gene 75 kb upstream of the C_α gene, with at least 2 D_δ regions and 2 J_δ regions arranged linearly 5' to C_δ (5'-$D_\delta 1$-$D_\delta 2$-$J_\delta 1$-$J_\delta 2$-C_δ-3') (Chien *et al.*, 1987a; b) $J_\delta 1$ lies 12.8 kb upstream of C_δ while $J_\delta 2$ is 5.6 kb upstream of C_δ. The 11-bp $D_\delta 1$ is 9.8 kb 5' to $J_\delta 1$, while the 16-bp $D_\delta 2$ lies only 900 bp upstream of $J_\delta 1$. In humans, the δ locus is very similar (see Fig. 8) although only one D region has been isolated to date. The human C_δ begins 85 kb upstream of Cα (Champagne *et al.*, 1988; Griesser *et al.*, 1988; Takihara *et la.*, 1988a). There are four C_δ exons. The first encodes the majority of the extracellular constant-region domain, while the second codes for a hinge-like region and the third encodes the transmembrane and cytoplasmic portions. The last exon contains only 3' untranslated sequences (Takihara *et al.*, 1988b). Both the amino acid sequence and the exon arrangement of the C_α and C_δ genes are very similar, suggesting that they arose by a gene duplication event. The 16bp $J_\delta 1$ and the 17bp $J_\delta 2$ are located 5.7 and 12 kb upstream of the 5' most exon of C_δ respectively (Griesser *et al.*, 1988; Takihara *et al.*, 1988b). A functional D segment, $D_\delta 1$, is found 1 kb upstream of J1. A screening of the region 40 kb upstream of $J_\delta 1$ reveals no crosshybridization to any D or J region probes, suggesting that there are only two J_δ and one D_δ in humans (Takihara *et al.*, 1988b). Screening of this same region with known V_α genes reveals no crosshybridization, raising the possiblity of V_δ in this region (Griesser *et al.*, 1988).

Rearrangement of T-cell receptor gene segments is governed by flanking recombinational signals that consist of a highly conserved heptamer, proximal to the gene segment, separated from a conserved A/ T rich nonamer by a non-conserved spacer. The spacer can either be long (~23 bp or two turns of the DNA helix) or short (~12 bp or one turn of the DNA helix). The distributions of spacers determines which segments can recombine, since only signals with long spacers recombine with those that have short spacers and vice versa. There are short spacers 5' to both the D_δ and J_δ gene segments and long spacers 3' to the D_δ genes and V segments (Chien *et al.*, 1976b; Takihara *et al.*, 1988b). This

Figure 8. A restriction map of the human δ-chain locus, using the enzymes Eco RI (E), Bam HI (B), and Hind III (H). There is one diversity element ($D_\delta 1$) and two joining gene segments, $J_\delta 1$ and J$\delta 2$. The constant region gene consists of four exons, the first three of which are coding exons (shading boxes), while the fourth (open box) codes for the 3' untranslated region of the message.

arrangement of spacers raised a number of interesting possibilities. Since the long and short spacers of D_δ segments are adjacent, there can be $D_\delta D_\delta$ joining, creating a VDDJ rearrangement such as has been found by Chien *et al.* (1987b). The presence of a short spacer 5' to the J_δ regions and long spacers 3' to the V_δ genes allows for the optional use of the D_δ segments. The same distribution of spacers adjacent to the V genes (long spacers) and J genes (short spacers) of both the α-and δ- chain genes (see Fig. 9), raises the possibility that the gene pools of the two chains may be shared. Comparison of δ- and α- chain cDNA sequences may clarify this question. To date, no V_δ gene has been isolated that is identical to a known V_α, although two murine V_δs are highly homologous to certain V_α (Chien *et al.*, 1987a, b). At least seven different murine V_δ have been isolated (Chien *et al.*, 1987a, b; D. Raulet, personal communication), which bear little resemblance to V_α genes.

The extent of the γδ repertoire seems to be more limited than that of the αβ receptor. Analysis of γδ+ thymocytes with γ- and δ-chain probes reveals sev-

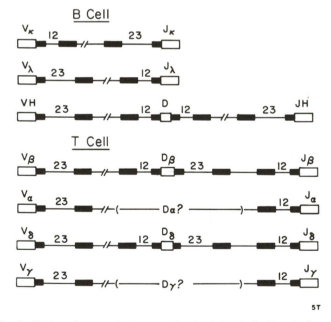

5T

Figure 9. The distribution of spacers in the recombinational signals flanking the immunoglobulin and T-cell-receptor genes. In immunoglobulin heavy-chain genes the D region is necessary, as the one turn/two turn rule prohibits direct VJ joining, but in both the β and δ loci, use of the D segments is optional. The presence of long spacers 3' to both the V_α and V_δ genes raises the possibility that the two chains may share V genes, but there is no evidence to support this hypothesis yet.

eral predominant rearrangements at each locus, suggesting that relatively few V_γ and V_δ genes are expressed by the majority of $\gamma\delta+$ thymocytes (D. Raulet, personal communication). It is not clear whether this limited repertoire is simply the result of a paucity of V genes or whether structural constraints on $\gamma\delta$ chain pairing and selection of cells with certain V genes play a significant role in the generation of a limited repertoire. It is also possible that the use of only a few V_δ genes is only characteristic of cells at this stage of development and not of $\gamma\delta+$ cells in the periphery.

The unique position of the δ locus within the α locus would result in the deletion of δ-chain genes in the course of α chain rearrangement, which, presuming allelic exclusion, would preclude concurrent expression of functional α- and δ-chain genes with different specificities in a given T cell. This does not, however, preclude the production of a δ-chain protein and an α-chain protein with the same variable domain in the same cell by differential splicing (see Fig. 10).

Rearrangement of the δ-chain genes occurs early in ontogeny, long before C_α containing messages or proteins can be detected. In hybridomas with day-14 fetal thymocytes, δ rearrangements predominantly involve the $J_\delta 1$ gene, with a mixture of complete (VDJ) and incomplete (DJ) rearrangements, and a large proportion of these cells have β- or γ-chain rearrangements as well (Chien *et al.*, 1987b). Most of day-15 hybridomas examined had rearranged δ, and of those that had rearranged γ, most had $C_\gamma 1$ rearrangements in conjunction with potentially functional δ rearrangements. On day 16, however, the picture

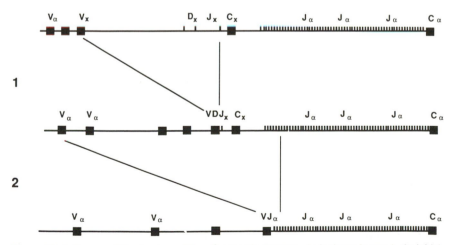

Figure 10. A diagram of the structure of the α/δ genomic structure. As is shown in step 1, the initial rearrangement can juxtapose a variable region with a J_δ gene segment. This could be transcribed and then produce either a message containing the C_δ gene or, if the primary transcript includes the C_α gene, a message. A second rearrangement can then occur to produce an α chain gene (step 2).

changes. Of the day-16 hybridomas that rearranged γ, most used $C_\gamma 1$ genes accompanied by a δ locus in either the germline form or a nonfunctional δ rearrangement, bearing out the hypothesis that surface $\gamma\delta$ expression is found early in T-cell ontogeny and is for the most part $C_\delta 2$ derived. By day 18, complete δ rearrangement (VDJ) or deletion of the δ locus has occurred on both chromosomes of the hybridomas tested. A small percentage of day-18 hybridomas had α-chain rearrangements (Chien *et al.*, 1987b), consistent with the appearance of surface $\alpha\beta$ receptor on thymocytes at this time. An analysis of day-18 and -19 fetal thymocyte genomic DNA revealed $D_\delta D_\delta J_\delta$, $V_\delta D_\delta$, and $V_\delta D_\delta J_\delta$ rearrangements along with a reciprocal joining of reciprocal recognition signals, resulting in head-to-head positioning of heptamers (Chien *et al.*, 1987b), indicating that mechanisms other than looping out/excision are being used in δ-chain rearrangement.

5. Possible Function of the $\gamma\delta$ Receptor

The cytolytic activity of the human $\gamma+$ T lymphocytes studied by Brenner *et al.* (1987) does not seem to be MHC-restricted, suggesting the possibility that recognition by the γ/X receptor is very different from that of the classical $\alpha\beta$ T-cell receptor. Analysis of CD3+ NK cells has shown that they bear a γ-chain complex on their cell surface which may be involved in their target recognition (Moingeon *et al.*, 1986; Moingeon *et al.*, 1987). Analysis of cells derived from *in vitro* culture of $\gamma+\delta+$ cells has shown that these cells are Fc receptor+ and that they exhibit antibody-dependent cellular cytotoxicity (ADCC), in addition to non-MHC restricted killing (Bank *et al.*, 1987; Borst *et al.*, 1987; Brenner *et al.*, 1987). These cells can also be activated by the combination of IL-2 and γ-interferon. Matis *et al.* (1987) have put forward an explanation for the non-MHC restricted cytotoxicity. They have isolated $\gamma\delta$ + cells that exhibit only MHC-restricted killing, which upon culture with exogenous IL-2 acquire the ability to kill in a non-MHC-restricted fashion. Since all the *in vitro* cultures that exhibit the nonrestricted killing had exogenous IL-2 added and since CTL cultures can be induced to exhibit nonspecific killing upon addition of high levels of IL-2, Matis *et al.* (1987) propose that the lack of restriction is an artifact of the IL-2 addition and that $\gamma\delta$ + cells *in vivo* are MHC-restricted. It is clear that further experiments will have to be performed to address this allegation. A summary of the possible roles of γ/δ receptor may play is given in Table II.

6. Conclusion

Investigations of the structure and function of the enigmatic $\gamma\delta$ receptor have often raised more questions than they have answered, but these new questions have served to expand our thinking regarding the role that T-cell recogni-

Table II. Possible Role(s) of Second Receptor (g-X)

1. Two receptors (a, MHC; b, Ag)	Unlikely
2. Precursor receptor (transfer specificity through γ-β, e.g., γ-β—αβ)	Unlikely
3. Anti-self receptor (high in autoreactive)	Possible
4. Natural killer receptor	Possible
5. Alternate chain (like λ and k)	Unlikely
6. Primitive (historically) receptor for first line defense	Likely
7. Recognize novel class I-like molecules	Likely

tion plays in the immune system and our concepts of T-cell ontogeny and selection. It is becoming more and more clear that the γδ dimer is an important cell-surface molecule on certain select T-cell subsets, and even on such non-T cell lymphocytes as NK cells; however the role that it plays in their target recognition, if any, still remains to be deduced.

References

Bank, I., DePinho, R.A., Brenner, M.B., Cassimeris, J., Alt., F.A., and Chess, L., 1986, A functional CD3 molecule associated with a novel heterodimer on the surface of immature human thymocytes, *Nature* **322:**179–181.

Bluestone, J.A., Pardoll, D., Sharrow, S., and Fowlkes, B.J., 1987, Characterization of murine thymocytes with CD3-associated T-cell structures, *Nature* **326:**82–84.

Born, W., Rathburn, G., Tucker, P., Marrack, P., and Kappler, J., 1986, Synchronized rearrangement of T cell γ and β chain genes during fetal thymocyte development, *Science* **234:**479–482.

Born, W., Yague, J., Palmet, E., Kappler, J., and Marrack, P., 1985, Rearrangement of T-cell receptor β-chain genes during T cell development, *Proc. Natl. Acad. Sci. USA* **82:**2925–2929.

Born, W, Miles, C., White, J., O'Brien, R., Freed, J.H., Marrack, P., Kappler, J., and Kubo, R.T., 1987, Peptide sequences of the T-cell receptor δ and γ chains are identical to predicted x and δ proteins, *Nature* **330:**572–574.

Borst, J., Alexander, S., Elder, J., and Terhorst, C., 1983, The T3 complex on human T lymphocytes involves four structurally distinct glycoproteins, *J. Biol. Chem.* **258:**5135–5143.

Borst, J., van de Grund, R.J., van Osteen, J.W., Ang, S.L., Melief, C.J., Seidman, J.G., and Bolhuis, R.L.H., 1987, A T cell receptor gamma/CD3 complex on CD3+4-8- cloned cytolytic peripheral blood lymphocytes, *Nature* **325:**683–688.

Boryhadi, M., Weiss, A., Tucker, P.W., Tigelaar, R.E., Allison, J.P., 1987, Delta is the Cₓ-gene product in the γ/δ antigen receptor in dendritic epidermal cells, *Nature* **330:**574–576.

Brenner, M.B., MacLean, J., Dialynas, D.P., Strominger, J.L., Smith, J.A., Owen, F.L., Seidman, J.G., Ip, S., Rosen, F., and Krangel, M.S., 1986, Identification of a putative second T-cell receptor, *Nature* **322:**145–149.

Brenner, B., McLean, J., Scheft, H., Riberdy, J., Ang, S., Siedman, J., Devlin, P., and Krangel, M.S., 1987, Two forms of the T cell receptor γ protein found on peripheral blood cytotoxic T lymphocytes, *Nature* **325:**689–693.

Budd, R.C., Cerottini, J.C., and McDonald, H.R., 1986, Cultured Lyt2-L3T4- T lymphocytes from normal thymus of mice express a broad spectrum of cytolytic activity, *J. Imm.* **137:**3734–3741.

Champagne, E., Takihara, Y., Sagman, U., Griesser, H., Tkachuk, D., Isuza, J., Mak, T.W., and

Minden, M.D., 1987, The isolation of the human δ chain gene and its expression in normal T cells and T cell leukemias, (submitted).

Chien, Y., Becker, D., Lindsten, T., Okamura, M., Cohen, D., and Davis, M., 1984a, A third type of murine T cell receptor gene, *Nature* **312:**31–35.

Chien, Y.-H., Gascoigne, N.R.J., Kaveler, J., Lee, N.E., and David, M.M., 1984b, Somatic recombination in a murine T-cell receptor gene, *Nature* **309:**322–326.

Chien, Y.-H., Iwashima, M., Kaplan, K.B., Elliot, J.F., and Davis, M.M., 1987a, A new T-cell receptor gene located in the alpha locus and expressed early in T cell differentiation, *Nature* **327:**677–681.

Chien, Y.-H., Iwashima, M., Wittstein, D.A., Kaplan, K.B., Elliot, J.F., Born, W., and Davis, M.M., 1987b. T-cell receptor δ gene rearrangements in early thymocytes, *Nature* **330:**722–727.

Davies, D.R., and Metzger, H., 1983, Structural basis of antibody function, *Ann. Rev. Immunol.* **1:**63–86.

DelaHera, A., Toribo, M.L., Minguez, C., and Martinez, C., 1985, Interleukin-2 promotes growth and cytolytic activity in human T3⁻4⁻8⁻ thymocytes, *Proc. Natl. Acad. Sci. USA* **82:**6268–6271.

Dialynas, D.P., Murre, C., Quetermous, T., Boss, J., Leiden, J.M., Seidman, J.G., and Strominger, J.L., 1986, Cloning and sequence analysis of complementary DNA encoding an aberrantly rearranged human T cell γ chain, *Proc. Natl. Acad. Sci. USA* **83:**2619–2613.

Garman, R.D., Doherty, P.J., and Raulet, D.H., 1986, Diversity, rearrangement and expression of murine T cell gamma genes, *Cell* **45:**733–742.

Griesser, H., Champagne, E., Tkachok, D., Takihara, P., Lalande, M., Baillie, E., Minden, M.D., and Mak, T.W., 1987, Mapping of the human T cell receptor α-δ region: A locus with a new constant region gene and prone to multiple chromosomal translocations, (submitted).

Haars, R., Kronenberg, M., Owen, F., Gallatin, M., Weissman, I., and Hood, L., 1986, Rearrangement and expression of T-cell antigen receptor and γ chain genes during thymic differentiation, *J. Exp. Med.* **164:**1–24.

Haskins, K., Kappler, J., and Marrack, P., 1984, The major histocompatibility complex-restricted antigen receptor on T cells, *Ann. Rev. Immunol.* **2:**51–66.

Hayday, A.C., Saito, H., Gillies, S.D., Kranz, D.M., Tanigawa, G., Eisen, H.N., and Tonegawa, S., 1985, Structure, organization and somatic rearrangement of T cell γ genes, *Cell* **40:**259–269.

Hedrick, S.M., Cohen, D.I., Nielsen, E.A., and Davis, 1984a, Isolation of cDNA clones encoding T cell specific membrane-associated proteins, *Nature* **308:**149–152.

Hedrick, S.M., Nielsen, E.A., Kavaler, J., Cohen, D.I., Davis, M.M., 1984b, Sequence relationships between putative T-cell receptor polypeptides and immunoglobulins, *Nature* **308:**153–158.

Helig, J.S., and Tonegawa, S., 1986, Diversity of murine γ genes and expression in fetal and adult thymocytes, *Nature* **322:**336–340.

Helig, J.S., Glimcher, L.H., Kranz, D.M., Clayton, L.K., Greenstein, J.L., Saito, H., Maxam, A.M., Burakoff, S.J., Eisen, H.N., and Tonegawa, S., 1985, Expression of the T-cell specific γ gene is unnecessary in T cells recognizing class II MHC determinants, *Nature* **317:**68–70.

Honjo, T., 1983, Immunoglobulin genes, *Annu. Rev. Immunol.* **1:**499–528.

Iwamoto, I., Ohashi, P., Walker, C., Rupp, F., Yoho, H., Hengartner, H., and Mak, T.W., 1986, The murine γ chain genes in B10 mice: Sequence and expression of new constant and variable genes, *J. Exp. Med.* **163:**1203–1212.

Jones, B., Mjolsness, S., Janeway, C., and Hayday, A.C., 1986, Transcripts of functionally rearranged gamma genes in primary T cells of immunocompetent mice, *Nature* **323:**635–638.

Kappler, J., Kubo, R., Haskins, K., White, J., and Marrack, P., 1983, The mouse T-cell receptor: Comparison of MHC-restricted receptors on two T-cell hybridomas, *Cell* **34:**727–737.

Kimura, N., Du, R.P., and Mak, T.W., Rearrangement and organization of T cell receptor gamma chain genes in human leukemic T cell lines, *Eur. J. Immunol.* (in press).

Kishihara, K., Yoshikai, Y., Matsuzakai, G., Mak, T.W., and Nomoto, K., 1987, Functional α and β chain T cell receptor messages can be detected in old, but not young, athymic mice, *Eur. J. Immunol.* **17**:477–482.

Krangel, M.S., Band, H., Hata, S., McLean, J., and Brenner, M.B., 1987, Structurally divergent human T cell receptor γ proteins encoded by distinct $C_γ$ genes, *Science* **237**:64–67.

Kranz, D.M., Saito, H., Disteche, C.M., Swisshelm, I., Pravtcheva, D., Ruddle, F., Eisen, H.N., and Tonegawa, J., 1985a, Chromosomal locations of the T-cell receptor α chain gene and the T cell α gene, *Science* **227**:941–945.

Kranz, D.M., Saito, H., Heller, M., Takagaki, Y., Haas, W., Eisen, H., and Tonegawa, S., 1985b, Limited diversity of the rearranged T cell γ gene, *Nature* **313**:752–755.

Lanier, L.L., Federspiel, N.A., Ruitenberg, J.J., Phillips, J.H., Allison, J.A., Littman, D., and Weiss, A., 1987, The T cell receptor complex expressed on normal peripheral blood CD4⁻CD8⁻ T lymphocytes: A CD3-associated disulphide-linked γ chain heterodimer, *J. Exp. Med.* **165**:1067–1094.

Lanier, L.L., and Weiss, A., 1986, Presence of Ti(WT31) negative T lymphocytes in normal blood and thymus, *Nature* **324**:268–270.

LeFranc, M.F., and Rabbitts, T., 1985, Two tandemly organized human genes encoding the T cell γ constant region sequences show multiple rearrangements in different T cell types, *Nature* **316**:464–466.

LeFranc, M.F., Forster, A., and Rabbits, T.H., 1986a, Rearrangement of two distinct T cell γ chain genes variable region genes in human DNA, *Nature* **319**:420–422.

LeFranc, M.F., Forster, A., and Rabbits, T.H., 1986b, Genetic polymorphism and exon changes of the constant regions of the human T cell rearranging gene γ, *Proc. Natl. Acad. Sci. USA* **83**:9596–9600.

LeFranc, M.F., Forster, A., Baer, R., Stinson, M.A., and Rabbits, J., 1986c, Diversity and rearrangement of the human T cell rearranging γ genes: Nine germline variable genes belonging to two subgroups, *Cell* **45**:237–247.

LeFranc, M.F., Forster, A., and Rabbitts, T., 1987, Organization of the human T cell rearranging gamma genes (TRG γ), in: *The T cell receptor*, UCLA symposia on Molecular and Cellular Biology, Volume 73 (J. Kappler and M. Davis, eds.), Alan R. Liss, New York, (in press).

Lew, A.M., Pardoll, D.M., Maloy, W.L., Fowlkes, B.J., Kruisbeek, A., Cheng, S.F., Germain, R.H., Bluestone, J.A., Schwartz, R.H., and Coligan, J.E., 1986, Characterization of T cell receptor γ chain expression in a subset of murine thymocytes, *Science* **234**:1401–1405.

Littman, D.G., Newton, M., Crommie, D., Ang, S.-L., Seidman, J.G., Gettner, S.N., and Weiss, A., 1987, Characterization of an expressed CD3-associated Ti γ-chain reveals $C_γ$ chain polymorphism, *Nature* **326**:85–88.

Loh, E.Y., Lanier, l.L., Turck, C.W., Littman, D.R., Davis, M.M., Chien, Y.H., and Weiss, A., 1987, Identification and sequence of a fourth human T cell antigen receptor chain, *Nature* **330**:569–572.

Maeda, K., Makamishi, M., Rogers, B., Haser, W.G., Shitara, K.,. Yoshida, H., Takagaki, Y., Augustin, A.A., and Tonegawa, S., 1987, Expression of the T cell receptor γ-chain gene product on the surface of peripheral T cells and T cell blasts generated by allogeneic mixed lymphocyte reaction, *Proc. Natl. Acad. Sci. USA* **84**:6536–6540.

Matis, L.A., Cron, R., and Bluestone, J.A., 1987, Major histocompatibility complex-linked, specificity of γδ receptor-bearing T lymphocytes, *Nature* **330**:262–264.

McIntyre, B.W., and Allison, J.P., 1983, The mouse T cell receptor: structural heterogeneity of molecules of normal T cells defined by xenoantiserum, *Cell* **34**:739–746.

Miescher, G.C., Budd, R.C., Lees, R.K., and McDonald, H.R., 1967, Abnormal expression of T cell receptor genes in Lyt2⁻ L3T4⁻ lymphocytes of lpr mice: Comparison with normal immature thymocytes, *J. Immunol.* **138**:1959.

Meuer, S.C., Cooper, D.A., Hodgdon, J.C., Hussey, R.E., Fitzgerald, K.A., Schlossman, S., and

Reinherz, E.L., 1983, Identification of the receptor for antigen and major histocompatibility complex on human inducer T lymphocytes, *Science* **222:**1239–1242.

Moingeon, P., Ythier, A., Goubin, G., Faure, F., Nowill, A., Delmon, L., Rainand, M., Forestier, F., Daffos, F., Bohuon, C., and Hercend, T., 1986, A unique T-cell receptor complex expressed on human fetal lymphocytes displaying natural-killer-like activity, *Nature* **323:**638–640.

Moingeon, P., Jitsukawa, S., Faure, F., Troalen, F., Triebel, F., Graziani, M., Forestier, F., Bellet, D., Bohoun, C., and Hercend, T., 1987, A γ chain complex forms a functional receptor on cloned human lymphocytes with natural killer-like activity, *Nature* **325:**723–726.

Murre, C., Waldman, R.A., Morton, C.C., Bongiovanni, K.F., Waldman, T.A., Shows, T.B., and Seidman, J.G., 1985, Human γ chain genes are rearranged in leukemic T cells and map to the short arm of chromosome 7, *Nature* **316:**549–552.

Nakashini, N., Maech, K., Ko-Ichi, I., Heller, M., and Tonegawa, S., 1987, γ protein is expressed on murine fetal thymocytes as a disulphide linked heterodimer, *Nature* **325:**720–722.

Ohashi, P., Mak, T.W., Van den Elsen, P., Yanagi, Y., Yoshikai, Y., Calman, A.F., Terhorst, C., Stobo, J.D., and Weiss, A., 1985, Reconstitution of an active T3/T cell antigen receptor in human T cells by DNA transfer, *Nature* **316:**602–606.

Pardoll, D.M., Fowlkes, B.J., Bluestone, J.A., Kruisbek, A., Maloy, W.L., Coligan, J.E., and Schwartz, R.H., 1987, Differential expression of two distinct T cell receptors during T cell development, *Nature* **326:**79–81.

Raulet, D.H., Garman, R.D., Saito, H., and Tonegawa, S., 1985, Developmental regulation of T-cell receptor gene expression, *Nature* **314:**103–107.

Reilly, A.E.B., Kranz, D.M., Tonegawa, S., and Eisen, H.N., 1986, A functional γ gene formed from known γ gene segments is not necessary for antigen-specific responses of murine cytotoxic T lymphocytes, *Nature* **321:**878–881.

Rupp, F., Frech, G., Hengartner, H., Zinkernagel, R.M., and Joho, R., 1986, No functional γ-chain transcripts detected in an alloreactive cytotoxic T cell clone, *Nature* **321:**876–878.

Saito, H., Kranz, D.M., Takagaki, Y., Hayday, A., Eisen, H., and Tonegawa, S., 1984a, A third rearranged and expressed gene in a clone of cytotoxic T lymphocytes, *Nature* **312:**36–40.

Saito, H., Kranz, D.M., Takagaki, Y., Hayday, A., Eisen, H., and Tonegawa, S., 1984b, Complete primary structure of a heterodimeric T cell receptor deduced from cDNA sequences, *Nature* **309:**757–762.

Sim, G., Yague, J., Nelson, J., Marrack, P., Palmer, E., Augustin, A., and Kappler, J., 1984, Primary structure of human T cell receptor α chain, *Nature* **312:**771–775.

Snodgrass, H.R., Dembic, Z., Steinmetz, M., and von Boehmer, H., 1985a, Expression of T-cell antigen receptor genes during fetal development in the thymus, *Nature* **313:**232–233.

Snodgrass, H.R., Kisielow, P., Kiefer, M., Steinmetz, M., and von Boehmer, H., 1985b, Ontogeny of the T-cell antigen receptor within the thymus, *Nature* **313:**592–595.

Takihara, Y., Champagne, E., Griesser, H., Kimura, M., Tkachuk, D., Reiman, J., Okada, A., Alt, F., Chess, L., Minden, M., and Mak, T.W., 1988a, The sequence and organization of the human T cell δ chain gene, *Eur. J. Immunol* (in press)

Takihara, Y., Tkachuk, D., Michalopoulos, E., Champagne, E., Reimann, J., Minden, M., and Mak, T.W., 1988b, Sequence and organization of the diversity, joining and constant region genes of the human T-cell δ chain locus, (submitted).

Toribo, M.L., Martinez, C., Marcos, M.A.R., Marquez, C., Calabraro, E., and DelaHera, A., 1986, A role for T3+4−6−8− transitional thymocytes in the differentiation of mature and functional T cells from human prothymocytes, *Proc. Natl. Acad. Sci. USA* **82:**6985–6988.

Weiss, A., and Stobo, J.D., 1984, Requirement for the coexpression of T3 and the T cell antigen receptor on a malignant T cell line, *J. Exp. Med.* **160:**1284–1299.

Weiss, A., Newton, M., and Crommie, D., 1986a, Expression of T3 in association with a molecule distinct from the T cell antigen receptor heterodimer, *Proc. Natl. Acad. Sci. USA* **83:**6998–7002.

Weiss, A., Imboden, J., Hardy, K., Manger, B., Terhorst, C., and Stobo, J., 1986b, The role of the CD3/antigen receptor complex in T cell activation, *Annu. Rev. Immunol.* **4**:593–619.

Yanagi, Y., Chan, A., Chin, B., Minden, M., and Mak, T.W., 1985, Analysis of cDNA clones specific for human T cells and the α and β chain of the T cell receptor heterodimer from a human T cell line, *Proc. Natl. Acad. Sci. USA* **82**:3430–3434.

Yanagi, Y., Yoshikai, Y., Leggett, K., Clark, S., Aleksander, I., and Mak, T.W., 1984, A human T cell-specific cDNA clone encodes a protein having extensive homology to immunoglobulin chains, *Nature* **308**:145–149.

Yoshikai, Y., Reis, M.D., and Mak, T.W., 1986a, Athymic mice express a high level of functional γ chain, but drastically reduced levels of α and β chain T cell receptor messages, *Nature* **324**:482–485.

Yoshikai, Y., Toyonaga, B., Koga, Y., Kimura, N., Griesser, H., and Mak, T.W., 1987, Repertoire of the human T cell γ gene: high frequency of non-functional transcripts in thymus and mature T cells, *Eur. J. Imm.* **17**:119–126.

Zauderer, M., Iwamoto, I., and Mak, T.W., 1986, Gamma gene expression in autoreactive helper T cells, *J. Exp. Med.* **163**:1314–1318.

Zinkernagel, R.M., Doherty, P.C., 1974, Restriction of in vitro T cell-mediated cytotoxicity in lymphocytic choriomeningitis within a syngeneic or semiallogeneic system, *Nature* **248**:701–702.

Index

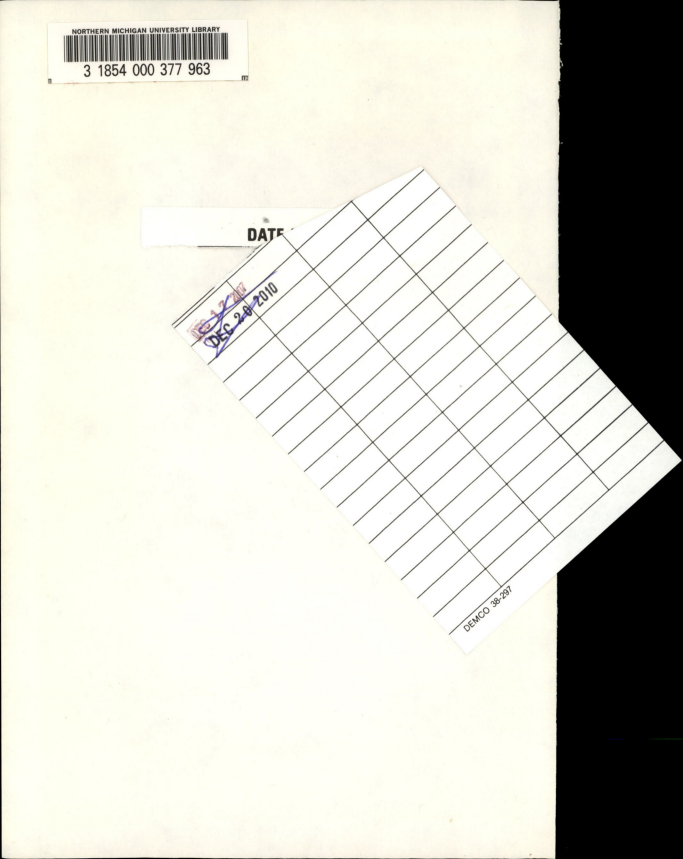

DATE

DEC 2 0 2010

DEMCO 38-297